REFERENCE NEUTRON RADIOGRAPHS
of nuclear reactor fuel

NEUTRONOGRAMMES DE REFERENCE
pour le combustible nucléaire

NEUTRONOGRAMMES DE REFERENCE
POUR
LE COMBUSTIBLE NUCLEAIRE

EUR 8916 EN EP   1984

Résumé

Faisant suite aux différentes publications du Groupe de Travail Européen sur la Neutronographie (NRWG), - constitué par des experts des centres de recherche nucléaire
des pays membres de la Communauté Européenne - , cet ouvrage rassemble l'expéri-
ence des principaux laboratoires de la Communauté Européenne dans le domaine du
contrôle de combustible nucléaire par neutronographie, en présentant un ensemble de
158 clichés de référence qui illustre les détails révélés par cette méthode sur des élé-
ments de réacteurs à eau légère et à neutrons rapides.
Une classification à double entrée de ces détails, complétée par une terminologie en six
langues, permet une consultation aisée des clichés, présentés à la fois sur films (docu-
ments pouvant être extraits pour examen comparatif, à l'échelle 1) et sur papier photographique (échelle 2), avec l'indication de l'origine du cliché et de la technique
employée.
Le recueil est complété par un rappel des principales disponibilités à l'intérieur de la
Communauté Européene en matière d'installations de neutronographie.

REFERENCE NEUTRON RADIOGRAPHS
OF
NUCLEAR REACTOR FUEL

EUR 8916 EN EP   1984

Abstract

This collection has been selected by a sub-group of the Neutron Radiography Working
Group (NRWG) - constituted by experts from nuclear research centres within the European Community - . The text is in English and French.
It contains 158 radiographs on film (which can be removed for comparative viewing) of
light water and fast reactor fuel rods. Also included are double enlargement prints of
each radiograph. The radiographs are arranged in accordance with a classification system, which lists the components of the fuel and cladding and illustrates defects in each
which can be revealed by neutron radiography.
Introductory chapters describe the classification system in relation to a typical fuel pin
and explain how to identify defects and use the collection. All the terms used in the classification together with additional useful terms are listed with their equivalents in six lan-
guages - Danish, Dutch, English, French, German and Italian.
A final section lists the main technical data and addresses of neutron radiography installations in the European Community.

Commission of the European Communities
Commission des Communautés Européennes
Joint Research Centre, Petten Establishment
Centre Commun de Recherche, Etablissement de Petten
Neutron Radiography Working Group (NRWG)
Groupe de Travail sur la Neutronographie

REFERENCE NEUTRON RADIOGRAPHS
of nuclear reactor fuel
NEUTRONOGRAMMES DE REFERENCE
pour le combustible
nucléaire

Edited by
Réalisé sous la direction de

J.C. DOMANUS - Risø

With contributions from
Avec la contribution de

R. Barbalat - Saclay
J.C. Domanus - Risø
J.F.W. Markgraf - Petten
F. Michel - Grenoble
D.J. Taylor - Harwell

D. REIDEL PUBLISHING COMPANY
A MEMBER OF THE KLUWER ACADEMIC PUBLISHERS GROUP
DORDRECHT / BOSTON / LANCASTER

Library of Congress Cataloging in Publication Data
Main entry under title:
Reference neutron radiographs of nuclear reactor fuel.

English and french.
At head of title: Commission of the European Communities=Communautés
Européennes. Neutron Radiography Working Group=Groupe de Travail sur la Neutronographie.
1. Nuclear fuels—Radiography. 2. Neutron radiography. I. Domanus, J. C.
II. Commission of the european communities. III. Euratom. Neutron radiography
working group. IV. Title: Neutronogrammes de reference pour le combustible
nucléaire.
TK9360.R44   1984   621.48'335   83-24777
ISBN-13: 978-94-009-6339-9   e-ISBN-13: 978-94-009-6337-5
DOI: 10.1007/ 978-94-009-6337-5

Publication arrangements by
Commission of the European Communities
Directorate-General Information Market and Innovation, Luxembourg

Design frontcover: J. Wells, ISPRA
Lay-out: Reproduction service J. R. C. PETTEN
Copies of neutron radiographs: P. Nielsen RISØ NATIONAL LABORATORY

EUR 8916 EN EP
© 1984 ECSC, EEC, EAEC, Brussels and Luxembourg
Softcover reprint of the hardcover 1st edition 1984

Published by D. Reidel Publishing Company
P.O. Box 17, 3300 AA Dordrecht, Holland

Sold and distributed in the U.S.A. and Canada
by Kluwer Academic Publishers,
190 Old Derby Street, Hingham, MA 02043, U.S.A.

In all other countries, sold and distributed
by Kluwer Academic Publishers Group,
P.O. Box 322, 3300 AH Dordrecht, Holland

TABLE DES MATIERES

TABLE OF CONTENTS

PREFACE

Although the principles of radiography with neutron beams have been known for some 45 years, their practical application in industry and research is still a rather young field. Norms, standards, and common terms of reference are scarce. One of the main tasks of the Neutron Radiography Working Group (NRWG) -constituted by the Joint Research Centre Petten of the Commission of the European Communities and national nuclear research centres within the European Community - has been to fill this gap.

It's common efforts have already resulted in two major publications

- the Neutron Radiography Handbook of November 1981 (EUR 7622e),
- the Proceedings of the First World Conference on Neutron Radiography, co-edited with American experts in March 1983 (EUR 8296 EN).

Together with this neutron radiograph reference collection they witness the frank and competent collaboration of European experts in the field.

P. von der HARDT
Commission of the European Communities
Joint Research Centre
Petten Establishment

PREFACE

Quoique les principes de la radiographie par rayons neutroniques soient connus depuis environ 45 ans, leur application pratique dans l'industrie et la recherche est encore un domaine plutôt récent.
Les normes, standards et termes de référence courants sont rares. Une des principales tâches du Groupe de Travail Européen sur la Neutronographie (NRWG) - constitué par des experts du Centre Commun de Recherche Petten de la Commission des Communautés Européennes et par des experts des centres de recherche nucléaire des pays membres de la Communauté Européenne - a consisté a combler cette lacune.

Ses efforts communs ont déjà eu pour résultat deux publications majeures:

o Le Manuel de Neutronographie de novembre 1981 (EUR 7622 e);
o Les débats de la 1ère Conférence Mondiale sur la Neutronographie, édités conjointement avec les experts américains en mars 1983 (EUR 8296 EN).

Joints à cette édition du recueil de neutronogrammes de référence, ils témoignent de la collaboration franche et compétente des experts européens dans ce domaine.

P. von der HARDT
Commission des Communautés Européennes
Centre Commun de Recherche
Etablissement de Petten

1. INTRODUCTION

Un Groupe de Travail sur la Neutronographie (NRWG) a été constitué en 1979 dans le cadre de la Communauté Européenne. Le Laboratoire National de Risø a publié en 1979 une contribution à ce groupe de travail sous forme d'un atlas intitulé *Détails révélés par la Neutronographie sur du Combustible de la filière à eau légère* [1]. Dans cet atlas, une classification de défauts typiques décelables sur ce type de combustible par neutronographie était établie. A partir de quelque 2000 clichés réalisés à Risø depuis 1979 par neutronographie, 36 avaient été choisis pour illustrer les défauts * portés dans cette classification.

L'une des principales tâches du programme du NRWG était de publier une collection plus complète de tels radiogrammes obtenus sur du combustible nucléaire, à partir de clichés fournis par tous les membres du NRWG.

Pour cela, un sous-groupe chargé de la nouvelle édition du recueil de neutronographie a été constitué avec la participation des personnes suivantes:

- R. Barbalat, Commissariat à l'Energie Atomique, IRDI/DERPE - Services des Piles de Saclay, Centre d'Etudes Nucléaires de Saclay, F - 91191 Gif-sur-Yvette Cédex (France), Tél: F-6-9082521 Telex: energ 690641 f
- J. Domanus (Président), Risø National Laboratory, Metallurgy Department, Postbox 49, DK - 4000 Roskilde (Danemark), Tél: DK-2-371212 Telex: ris 43116 dk
- J.F.W. Markgraf, Commission des Communautés Européennes, Centre Commun de Recherche, Division du RHF, Etablissement de Petten, Postbus 2, NL - 1755 ZG Petten (Pays-Bas), Tél: NL-2246-5146 Telex: reacp 57211 nl
- F. Michel, Commissariat à l'Energie Atomique, IRDI/DERPE Service des Piles de Grenoble, Centre d'Etudes Nucléaires de Grenoble, Boite Postale 85 X, F - 38041 Grenoble Cédex (France), Tel:F-76-974111 Telex: energ 320323 f
- D.J. Taylor, United Kingdom Atomic Energy Authority, Research Reactors Division, Bldg. 443, AERE, Harwell, Oxfordshire OX11 ORA (Grande-Bretagne), Tél:GB-235-24141 Telex: atomha 83135 uk

Les personnes mentionnées ci-dessus ont fourni les neutronogrammes contenus dans cette collection.

Certaines données techniques de l'ensemble des dispositifs de neutronographie dans les pays membres de la Communauté Européenne sont fournies au chapitre 8 ci-après.

---

*) Le terme "défaut" sert à désigner sur un cliché de neutronographie un détail présentant, à un certain niveau du combustible, après irradiation, une apparence différente de celle correspondant à un cliché pris à ce même niveau en l'état initial de fabrication.

1. INTRODUCTION

In 1979 a Neutron Radiography Working Group (NRWG) was constituted within the European Community. As a contribution to the NRWG, Risø National Laboratory published in June 1979, an atlas entitled: *Neutron Radiography Findings in Light Water Reactor Fuel* [1]. In this atlas a classification of typical defects which can be revealed by neutron radiography of light water reactor fuel was given. From about 2000 neutron radiographs taken up to 1979 at Risø, 36 neutron radiographs were chosen to represent the defects*) mentioned in the classification.

One of the main tasks on the NRWG program was the publication of a more complete collection of neutron radiographs of nuclear reactor fuel with radiographs originating from all NRWG members.

Therefore in 1982, a Sub-Group with the task of compiling the new edition of the reference neutron radiograph collection was constituted within NRWG with the participation of the following persons:

- R. Barbalat: Commissariat à l'Energie Atomique, IRDI/DERPE- Services des Piles de Saclay, Centre d'Etudes Nucléaires de Saclay,F-91191 Gif-sur-Yvette Cedex, France. Tel. F-6-9082521, telex energ 690641 f.
- J.C. Domanus (chairman): Risø National Laboratory, Metallurgy Department, Postbox 49, DK-4000 Roskilde, tel. DK-2-371212,telex risoe 43116 dk.
- J.F.W. Markgraf: Joint Research Centre of the Commission of the European Communities, Petten Establishment, HFR Division, Postbus 2, NL-1755 ZG Petten, The Netherlands, tel. NL-2246-5146, telex reacp 57211 nl.
- F. Michel: Commissariat à l'Energie Atomique, IRDI/DERPE - Service des Piles de Grenoble, Centre d'Etudes Nucléaires de Grenoble, F-38041 Grenoble Cedex, France. Tel. F-76-974111, telex energ 320323 f.
- D.J. Taylor: United Kingdom Atomic Energy Authority, Research Reactors Division, Bldg. 443, AERE Harwell, Oxfordshire OX11 ORA, United Kingdom, tel. GB-235-24141, telex atomha g 83135 uk.

The persons mentioned above have contributed all the neutron radiographs contained in this collection.

Some technical data on all the neutron radiographic facilities within the member states of the European Community are given in chapter 8 below.

---

*)The term "defect" is used to designate a change in appearance shown on an original neutron radiograph of a particular part of the fuel as fabricated, to that shown on a subsequent radiograph, usually post irradiation.

3

2. CRAYONS COMBUSTIBLES

Pour réaliser le présent recueil de neutronogrammes, quelques exemples typiques de crayons combustibles ont été choisis. Ils concernent du combustible nucléaire aussi bien sous forme de pastilles que vibro-compacté, ou de structure annulaire.

La figure 1 décrit schématiquement tous les composants d'un combustible, tels qu'ils sont mentionnés dans la classification ci-dessous, avec les repères suivants:

A.   COMBUSTIBLE
A.a.   Pastilles
A.b.   Combustible annulaire
A.c.   Jeu entre pastilles
A.d.   Evidement
A.e.   Combustible vibro-compacté
A.f.   Jeu combustible-gaine
A.g.   Colonne combustible
A.h.   Constitution du combustible

B.   GAINE

C.   VOLUME LIBRE
C.a.   Ressort
C.b.   Manchon du ressort
C.c.   Disque d'isolement
C.d.   Séparateur
C.e.   Espace colonne combustible - bouchon

D.   BOUCHONS
D.a.   Bouchon inférieur
D.b.   Bouchon supérieur

E.   INSTRUMENTATION
E.a.   Thermocouple
E.b.   Capteur de pression
E.c.   Jauge de diamètre
E.d.   Jauge de longueur
E.e.   Autre instrumentation

2. FUEL PINS

For the purpose of the present collection of neutron radiographs some typical examples of nuclear fuel pins were chosen. They represent pelletized, annular and vibro-compacted nuclear fuel.
Figure 1 gives schematic diagrams of all the fuel components, as mentioned in the classification below. They are marked in the following way:

A.   FUEL
A.a. Pellets
A.b. Annular fuel
A.c. Pellet-to-pellet-gap
A.d. Dishing
A.e. Vibro compacted fuel
A.f. Fuel-to-clad-gap
A.g. Fuel column
A.h. Fuel composition

B.   CLADDING

C.   PLENUM
C.a. Spring
C.b. Spring sleeve
C.c. Insulating disc
C.d. Spacer
C.e. Fuel column to plug distance

D.   PLUGS
D.a. Bottom plug
D.b. Top plug

E.   INSTRUMENTATION
E.a. Thermocouple
E.b. Pressure transducer
E.c. Diameter gauge
E.d. Length gauge
E.e. Other instrumentation

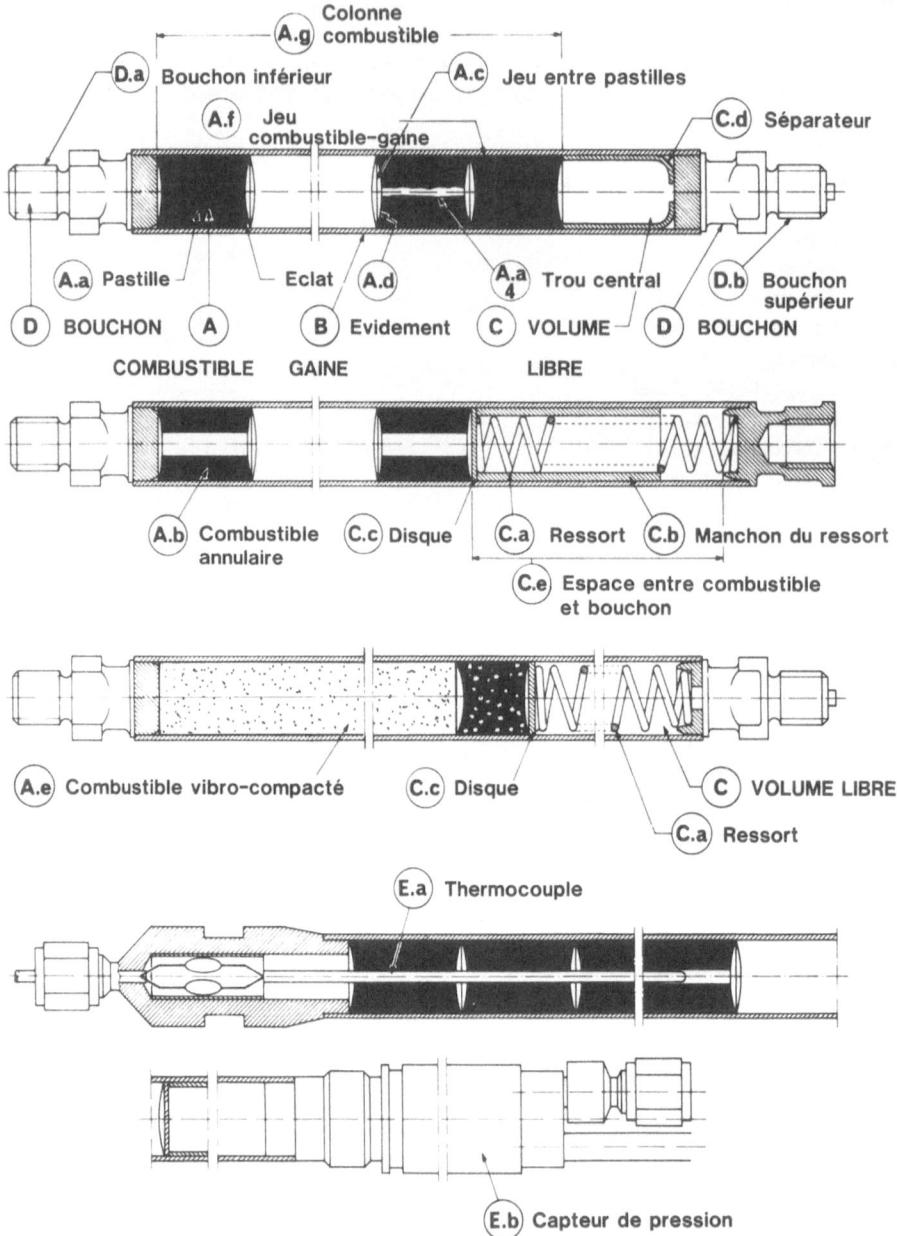

Fig. 1

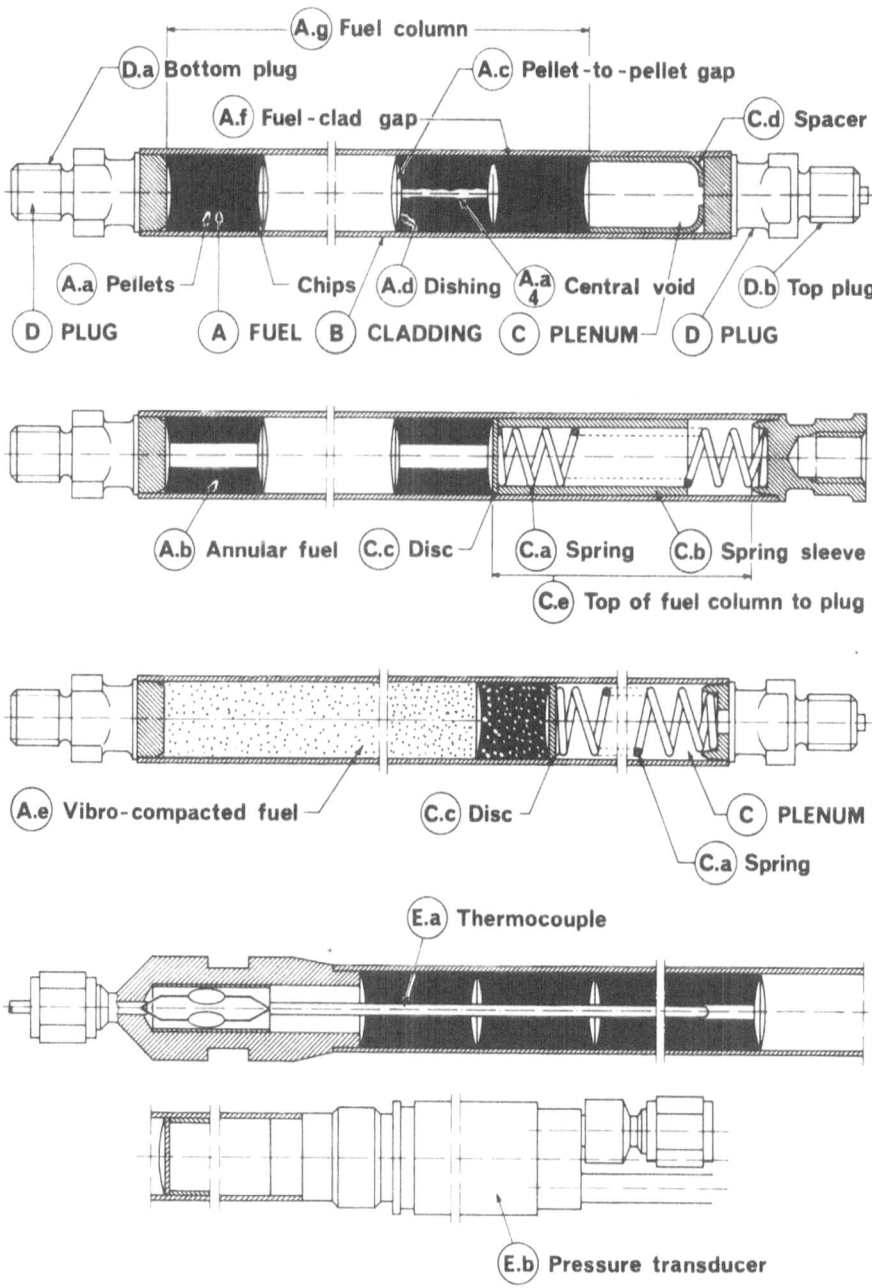

Fig.1

7

3. CLASSIFICATION DES DETAILS VUS PAR NEUTRONOGRAPHIE

Le tableau 1 présente la classification de différents détails vus sur un neutronogramme de combustible nucléaire. Sur le côté gauche du tableau se trouve une énumération des composants du crayon combustible (voir 2 ci-dessus). En haut du tableau et plus à droite sont énumérés tous les détails qui peuvent être révélés par neutronographie de crayons combustibles nucléaires.
Ces détails sont les suivants:

0. (Eléments de crayon combustible) A L'ETAT INITIAL
1. FISSSURES
1.1 Aléatoire
1.2 Longitudinal
1.3 Transversal
1.4 Circonférentiel
1.5 Stratifié

2. ECLATS
2.1 Angle
2.2 Autre
2.3 Dans jeu entre pastilles
2.4 Manquant

3. CHANGEMENT DE FORME OU DE PLACE
3.1 Elargi ou gonflé
3.2 Comprimé ou tassé
3.3 Comblé ou refermé
3.4 Déformé
3.5 Rompu
3.6 Déplacé ou disloqué
3.7 Allongé ou distendu
3.8 Accumulé
3.9 Restructuré
3.10 Fondu
3.11 Désintégré
3.12 Ayant migré

4. CAVITE CENTRALE
4.1 Dans une pastille
4.2 A travers plusieurs pastilles
4.3 A travers toute la colonne combustible

5. INCLUSIONS
5.1 De plutonium
5.2 De poison
5.3 Autre

6. CORROSION
6.1 Hydrures
6.2 Oxydes
6.3 Autre

7. QUALITES NEUTRONIQUES
7.1 Différence d'enrichissement
7.2 Différence de consommation

8. FLUIDE PRIMAIRE
8.1 Présent
8.2 Absent

Les détails énumérés ci-dessus sont illustrés par des exemples relatifs à des combustibles de la filière à eau légère (repères L) et de la filière à neutrons rapides (repères F). L'absence de repère L ou F dans le tableau signifie qu'aucun exemple n'est donné pour le cas correspondant. Comme l'existence d'un tel exemple est cependant possible, le directeur de cette publication serait reconnaissant à quiconque pourrait lui en fournir un, en vue de l'inclure dans la prochaine édition du recueil.

3. CLASSIFICATION OF NEUTRON RADIOGRAPHIC FINDINGS

Table 1 gives a classification of different findings on neutron radiographs of nuclear fuel. At the lefthand side of the table an enumeration of the fuel pin components is given (described in 2 preceding). At the top righthand side of the table all the findings which have been revealed by neutron radiography of nuclear fuel pins are listed. They are the following:

0.	(Fuel pin part) AS FABRICATED	4.	VOIDAGE
1.	CRACKS	4.1	In one pellet
1.1	Random	4.2	Through several pellets
1.2	Longitudinal	4.3	Through whole fuel column
1.3	Transverse		
1.4	Annular	5.	INCLUSIONS
1.5	Stratified	5.1	Of Plutonium
		5.2	Of poison
2.	CHIPS	5.3	Other
2.1	Corner		
2.2	Other	6.	CORROSION
2.3	In pellet-to-pellet gap	6.1	Hydrides
2.4	Missing	6.2	Oxides
		6.3	Other
3.	CHANGE OF SHAPE OR LOCATION		
3.1	Enlarged or swollen	7.	NUCLEAR PROPERTIES
3.2	Contracted	7.1	Different enrichment
3.3	Filled-up or closed	7.2	Different burn-up
3.4	Deformed		
3.5	Broken	8.	COOLANT
3.6	Dislocated	8.1	Present
3.7	Extended	8.2	Absent
3.8	Accumulated		
3.9	Restructured		
3.10	Melted		
3.11	Disintegrated		
3.12	Migrated		

The above listed findings are illustrated with examples from light water (marked - L) and fast reactor fuel (marked - F). If there is neither an L nor F marking in the table it means that no example of the corresponding finding is given. It is, however, possible that such an example exists and the editor of this collection will be grateful to receive a copy to be included in the next edition of the collection.

TABLEAU 1

COMBUSTIBLE DE LA FILIERE

A EAU LEGERE (L)
OU
A NEUTRONS RAPIDES (F)

Légende des colonnes (états/défauts) :

- 0 — A L'ÉTAT INITIAL
- 1-FISSURES : 1.1 ALÉATOIRE, 1.2 LONGITUDINAL, 1.3 TRANSVERSAL, 1.4 CIRCONFÉRENTIEL, 1.5 STRATIFIÉ
- 2-ÉCLATS : 2.1 ANGLE, 2.2 AUTRE, 2.3 DANS LE JEU ENTRE PASTILLES, 2.4 MANQUANT
- 3-CHANGEMENT DE FORME OU DE PLACE : 3.1 ÉLARGI OU GONFLE, 3.2 COMPRIMÉ OU TASSÉ, 3.3 COMBLE OU REFERME, 3.4 DÉFORMÉ, 3.5 ROMPU, 3.6 DÉPLACÉ OU DISLOQUÉ, 3.7 ALLONGE OU DISTENDU, 3.8 ACCUMULE, 3.9 RESTRUCTURE, 3.10 FONDU, 3.11 DÉSINTÈGRE, 3.12 AYANT MIGRE
- 4-CAVITE : 4.1 DANS UNE PASTILLE, 4.2 À TRAVERS PLUSIEUR PASTILLES, 4.3 À TRAVERS TOUTE LA COLONNE
- 5-INCLUSIONS : 5.1 DE PLUTONIUM, 5.2 DE POISON, 5.3 AUTRE
- 6-CORROSION : 6.1 HYDRURES, 6.2 OXYDES, 6.3 AUTRE
- QUALITES NEUTRON : 7.1 ENRICHISSEMENT, 7.2 CONSOMMATION
- 8-FLUIDE PRIM : 8.1 PRÉSENT, 8.2 ABSENT

Groupe / Composant	0	1.1	1.2	1.3	1.4	1.5	2.1	2.2	2.3	2.4	3.1	3.2	3.3	3.4	3.5	3.6	3.7	3.8	3.9	3.10	3.11	3.12	4.1	4.2	4.3	5.1	5.2	5.3	6.1	6.2	6.3	7.1	7.2	8.1	8.2
A COMBUSTIBLE — a PASTILLES	L/F	L/F	L/F	L	L		L	L	L	L	L				L	L		F	L/F	F	L/F	F	L	L	L/F	L	L					L/F		L	
A COMBUSTIBLE — b ANNULAIRE	L	L	L	L/F				L		F	F		F	L/F				F	F	F	L	F			F										
A COMBUSTIBLE — c JEU ENTRE PASTILLES	L										L																							L	
A COMBUSTIBLE — d EVIDEMMENT	L			L									L	L																					L
A COMBUSTIBLE — e VIBRO-COMPACTE	L									L	L														L										
A COMBUSTIBLE — f JEU GAINE-COMBUSTIBLE	L/F													F																					
A COMBUSTIBLE — g COLONNE COMBUSTIBLE	L/F											F		L/F			F			F	F	F													
A COMBUSTIBLE — h CONSTITUTION	L													L	L/F											L	L								
B GAINE — a GAINE	L											L		L							F								L					L	
C VOLUME LIBRE — a RESSORT	L														L	L																			
C VOLUME LIBRE — b MANCHON DU RESSORT	L														L																				
C VOLUME LIBRE — c DISQUE D'ISOLEMENT	L															L																			
C VOLUME LIBRE — d SEPARATEUR	L/F																												L						
D BOUCHONS — e DU COMBUST. AU BOUCHON	L																																	L	L
D BOUCHONS — a INFERIEUR	L														L														L					L	
D BOUCHONS — b SUPERIEUR	L/F							L																											
E INSTRUMENTATION — a THERMO-COUPLE	F																																		
E INSTRUMENTATION — b CAPTEUR DE PRESSION	L																																		
E INSTRUMENTATION — c JAUGE DE DIAMETRE	L																			L															
E INSTRUMENTATION — d JAUGE DE LONGEUR	L																																		
E INSTRUMENTATION — e AUTRE	L																																		

10

TABLE 1

L - LIGHT WATER FUEL
F - FAST REACTOR FUEL

Column code legend:

- 0 — AS FABRICATED
- 1 CRACKS: 1.1 RANDOM, 1.2 LONGITUDINAL, 1.3 TRANSVERSE, 1.4 ANNULAR, 1.5 STRATIFIED
- 2 CHIPS: 2.1 CORNER, 2.2 OTHER, 2.3 IN PELLET-TO-PELLET GAP, 2.4 MISSING
- 3 CHANGE OF SHAPE OR LOCATION: 3.1 ENLARGED OR SWOLLEN, 3.2 CONTRACTED, 3.3 FILLED-UP OR CLOSED, 3.4 DEFORMED, 3.5 BROKEN, 3.6 DISLOCATED, 3.7 EXTENDED, 3.8 ACCUMULATED, 3.9 RESTRUCTURED, 3.10 MELTED, 3.11 DISINTEGRATED, 3.12 MIGRATED
- 4 VOIDAGE: 4.1 IN ONE PELLET, 4.2 THROUGH SEVERAL PELLETS, 4.3 THROUGH WHOLE FUEL COLUMN
- 5 INCLUSIONS: 5.1 PLUTONIUM, 5.2 POISON, 5.3 OTHER
- 6 CORROSION: 6.1 HYDRIDES, 6.2 OXIDES, 6.3 OTHER
- 7 NUCLEAR PROPERTY: 7.1 ENRICHMENT, 7.2 BURN-UP
- 8 COOLANT: 8.1 PRESENT, 8.2 ABSENT

Group	Item	0	1.1	1.2	1.3	1.4	1.5	2.1	2.2	2.3	2.4	3.1	3.2	3.3	3.4	3.5	3.6	3.7	3.8	3.9	3.10	3.11	3.12	4.1	4.2	4.3	5.1	5.2	5.3	6.1	6.2	6.3	7.1	7.2	8.1	8.2
FUEL (A)	a PELLETS	L/F	L/F	L/F	L	L		L	L	L	L	L				L	L		F	L/F	F	L/F	F	L	L	L/F	L	L					L/F		L	
	b ANNULAR FUEL	L	L	L	L/F		F		L		F	F		F	L/F				F	F	F	L	F			F										
	c PELLET-TO-PELLET GAP	L										L																							L	L
	d DISHING	L									L			L	L																					
	e VIBRO-COMPACTED	L/F			L							L														L										
	f FUEL-TO-CLAD GAP	L/F											F		F			F																		
	g FUEL COLUMN	L/F													L/F						F	F														
	h FUEL COMPOSITION	L																					F				L	L								
CLAD (B)	a CLADDING	L													L	L/F		L				F								L						
PLENUM (C)	a SPRING	L											L		L		L																		L	
	b SPRING SLEEVE	L														L																				
	c INSULATING DISC	L														L	L																		L	
	d SPACER	L/F																			L															
	e FUEL COLUMN TO PLUG	L																																		L
PLUGS (D)	a BOTTOM	L																												L					L	
	b TOP	L/F																												L					L	
INSTRUMENTATION (E)	a THERMO-COUPLE	F																																		
	b PRESSURE TRANSDUCER	L							L							L																				
	c DIAMETER GAUGE	L																																		
	d LENGTH GAUGE	L																																		
	e OTHER	L																																		

4. REPERAGE DES NEUTRONOGRAMMES

Tous les cas repérés par *L, F* ou *L/F* sur le tableau 1 font l'objet d'une illustration dans la collection à l'aide d'un ou plusieurs exemples.

Les clichés sont reproduits à la fois sur film pour copie (dans la taille originale) et sur papier photographique (agrandis 2 fois).

Les agrandissements sur papier photographique portent les numéros d'ordre des clichés, tandis que les légendes accompagnent les reproductions sur film.

Chaque cliché comporte un repère pour identifier son origine, le type de combustible illustré et la technique d'exposition employée. Ce repérage (indiqué sur papier comme sur film) s'interprète de la façon suivante:
- la première lettre majuscule représente l'origine du cliché neutronographique (G: Grenoble, H: Harwell; P: Petten, R: Risø, S: Saclay);
- le nombre qui suit indique le numéro d'ordre du cliché dans la contribution du centre mentionné;
- la seconde lettre majuscule donne le type du combustible (L: eau légère, F: neutrons rapides);
- le dernier chiffre caractérise la technique d'exposition:
  1- méthode de transfert avec convertisseur In et film Kodak Industrex C;
  2- méthode de transfert avec convertisseur Dy et film Kodak Industrex SR;
  3- méthode de transfert avec convertisseur Dy et film Agfa-Gevaert Structurix D4;
  4- méthode directe avec convertisseur Gd et film Kodak Industrex SR;
  5- méthode directe avec film nitrocellulose Kodak-Pathé CN 85 B (convertisseur incorporé);
  6- méthode directe avec film nitrocellulose Kodak-Pathé CN 85 et convertisseur BN 1 (pour les clichés en provenance de Saclay, le film CN 85 a été utilisé aussi avec un converstisseur ^{6}LiF).

Toute information complémentaire relative à un cliché quelconque de la collection peut être obtenue auprès du centre ayant fourni la neutronographie (liste au chapitre 1 ci-dessus), en rappelant le repère associé au cliché concerné.

4. MARKINGS OF THE RADIOGRAPHS

All neutron radiographs marked with *L, F* or *L/F* on table 1 are reproduced in the collection with one or more examples.

The radiographs are reproduced both on duplicating film (in original size) as well as on photographic paper (twice enlarged).

The enlargements on photographic paper bear the consecutive numbers of the radiographs whereas all the captions are contained on the duplicating film.

Each radiograph has a marking which identifies the origin of the radiograph, the type of nuclear fuel and the exposure technique. This marking (visible both on paper and film) denotes the following:

- The first capital letter indicates the origin of the radiograph (G - Grenoble, H - Harwell, P - Petten, R - Risø, S - Saclay),
- the following number means the consecutive radiograph originating from the centre mentioned above.
- The second capital letter gives the type of reactor fuel (L - light water reactor, F - fast reactor)
- The last number refers to the exposure technique by which the neutron radiographs were produced:
  1 -transfer technique with In converter and Kodak Industrex C film,
  2 -transfer technique with Dy converter and Kodak Industrex SR film,
  3 -transfer technique with Dy converter and Agfa-Gevaert Structurix D4 film,
  4 -direct technique with Gd converter and Kodak Industrex SR film
  5 -Kodak-Pathé nitrocellulose CN 85 type B film,
  6 -Kodak-Pathé nitrocellulose CN 85 film with BN 1 converter (for radiographs from Saclay the CN 85 film was also used with the ^{6}LiF converter).

Further information on any radiographs in this collection can be obtained from the contributing neutron radiography centre (listed in section 1 above) by referring to the marking on the radiograph in question.

5. UTILISATION DU RECUEIL

Les copies sur film des neutronogrammes peuvent être examinées sans les extraire du recueil, en éclairant la page blanche placée derrière chaque film à l'aide d'une lampe de bureau à abat-jour. La page blanche ainsi éclairée sert de fond lumineux pour la lecture du cliché.

Les clichés de référence peuvent aussi être extraits de la collection et examinés sur table lumineuse conjointement avec la neutronographie d'actualité en cours d'étude.

5. HOW TO USE THE COLLECTION

The copies of the neutron radiographs on film can be viewed without removing them by illuminating the blank page which follows with a shaded desk lamp.

The reference radiograph may also be removed from the collection and be viewed on an illuminator together with the actual radiograph under assessment.

6. CONTENU DU RECUEIL

On trouvera ci-après une liste de tous les neutronogrammes regroupés dans le présent recueil. La liste comporte le numéro d'ordre du cliché, son repérage et sa description (d'après le tableau 1).

LISTE DE LA COLLECTION DE CLICHES NEUTRONOGRAPHIQUES

A. COMBUSTIBLE a. Pastilles 0. A l'état initial

1.	P1L4	A.a.0.
2.	P2F6	A.a.0.

A.COMBUSTIBLE a. Pastilles 1. Fissures

3.	H1L1	A.a.1.1.	Aléatoires
4.	S1L6	A.a.1.1.	Aléatoires
5.	G1F2	A.a.1.1.	Aléatoires
6.	S2F6	A.a.1.2.	Longitudinales
7.	P3L5	A.a.1.2.	Longitudinales
8.	S3L6	A.a.1.3.	Transversales
9.	P4L5	A.a.1.3.	Transversales
10.	R1L3	A.a.1.4.	Circonférentielles

A.COMBUSTIBLE a. Pastilles 2. Eclats

11.	S4L6	A.a.2.1.	D'angle
12.	R2L3	A.a.2.2.	Autre
13.	S5L6	A.a.2.2.	Autre
14.	R3L3	A.a.2.3.	Dans jeu entre pastilles
15.	S6L6	A.a.2.3.	Dans jeu entre pastilles
16.	P5L5	A.a.2.3.	Dans jeu entre pastilles
17.	R4L4	A.a.2.4.	Manquants
18.	P6L2	A.a.2.4.	Manquants (en angle)

A.COMBUSTIBLE a. Pastilles 3. Changement de forme ou de place

19.	R5L3	A.a.3.1.	Gonflement
20.	S7L6	A.a.3.1.	Gonflement
21.	R6L3	A.a.3.5.	Fracturation
22.	G2L2	A.a.3.5.	Fracturation (transversale)
23.	P7L2	A.a.3.5.	Fracturation
24.	P8L5	A.a.3.5.	Fracturation
25.	P9L2	A.a.3.6.	Dislocation
26.	G3F2	A.a.3.8.	Accumulation de combustible (hors de la gaine)
27.	G4F2	A.a.3.9.	Restructuration (après fusion, état antérieur en G5F2)
28.	P10L5	A.a.3.9.	Restructuration du combustible (avant)
29.	P11L5	A.a.3.9.	Restructuration du combustible (après)

6. CONTENTS OF THE COLLECTION

The following list identifies all the radiographs present in this collection. The list contains the consecutive number of the radiograph, its marking and descriptions (according to table 1).

LIST OF CONTENTS OF THE COLLECTION

A. Fuel a. Pellets 0. As fabricated

| 1. | P1L4 | A.a.0 |
| 2. | P2F6 | A.a.0. |

A. FUEL a. Pellets 1. Cracks

3.	H1L1	A.a.1.1. Random
4.	S1L6	A.a.1.1. Random
5.	G1F2	A.a.1.1. Random
6.	S2F6	A.a.1.2. Longitudinal
7.	P3L5	A.a.1.2. Longitudinal
8.	S3L6	A.a.1.3. Transverse
9.	P4L5	A.a.1.3. Transverse
10.	R1L3	A.a.1.4. Annular

A. FUEL a. Pellets 2. Chips

11.	S4L6	A.a.2.1. Corner
12.	R2L3	A.a.2.2. Other
13.	S5L6	A.a.2.2. Other
14.	R3L3	A.a.2.3. In pellet-to-pellet gap
15.	S6L6	A.a.2.3. In pellet-to-pellet gap
16.	P5L5	A.a.2.3. In pellet-to-pellet gap
17.	R4L4	A.a.2.4. Missing
18.	P6L2	A.a.2.4. Missing (corner)

A. FUEL a. Pellets 3. Change of shape or location

19.	R5L3	A.a.3.1. Enlarged or swollen
20.	S7L6	A.a.3.1. Enlarged or swollen
21.	R6L3	A.a.3.5. Broken
22.	G2L2	A.a.3.5. Broken (transversely)
23.	P7L2	A.a.3.5. Broken
24.	P8L5	A.a.3.5. Broken
25.	P9L2	A.a.3.6. Dislocated and disintegrated
26.	G3F2	A.a.3.8. Accumulated (fuel,outside of cladding)
27.	G4F2	A.a.3.9. Restructured (fuel,after melting, original as G5F2)
28.	P10L5	A.a.3.9. Restructured (fuel,before restructuring)
29.	P11L5	A.a.3.9. Restructured (fuel,after restructuring)

30.	G5F2	A.a.3.10.Melted (several pellets)
31.	P12F2	A.a.3.10.Melted
32.	G6F2	A.a.3.11.Disintegrated
33.	S8L6	A.a.3.11.Disintegrated
34.	P13L5	A.a.3.11.Disintegrated
35.	G7F2	A.a.3.12.Migration of fuel (outside of cladding)

A. FUEL a. Pellets 4. Voidage

36.	S9L6	A.a.4.1. Central void in one pellet
37.	S10L6	A.a.4.1. Central void in one pellet
38.	G8L2	A.a.4.2. Central void (filled up) through several pellets
39.	G9L2	A.a.4.2. Central void (enlarged) through several pellets
40.	S11L6	A.a.4.2. Central void through several pellets
41.	R7L3	A.a.4.3. Central void through fuel column
42.	P14F2	A.a.4.3. Central void through fuel column

A. FUEL a. Pellets 5. Inclusions

43.	S12L6	A.a.5.1. Of Pu
44.	S13L6	A.a.5.2. Of poison (left-granulated; right-powder)
45.	S14L6	A.a.5.2. Of poison (Gd_2O_3 powder, irradiated)

A. FUEL a. Pellets 7. Nuclear properties

46.	R8L3	A.a.7.1. Different enrichment
47.	S15L6	A.a.7.1. Different enrichment
48.	P15F2	A.a.7.1. Different enrichment

A. FUEL a. Pellets 8. Coolant

| 49. | G10L2 | A.a.8.1. In pellets |
| 50. | P16L5 | A.a.8.1. In pellets (at dishing) |

A. FUEL b. Annular 0. As fabricated

| 51. | P28L6 | A.b.0. |
| 52. | S20L6 | A.b.0. |

A. FUEL b. Annular 1. Cracks

53.	S21L6	A.b.1.1. Random
54.	S22L6	A.b.1.2. Longitudinal
55.	S23L6	A.b.1.3. Transverse
56.	G13F2	A.b.1.3. Transverse
57.	G14F2	A.b.1.5. Stratified

A. FUEL b. Annular 2. Chips

| 58. | S24L6 | A.b.2.2. Other |
| 59. | G15F2 | A.b.2.4. Missing |

A.COMBUSTIBLE b. Annulaire 3. Changement de forme ou de place

60.	G16F2	A.b.3.1.	Elargissement du trou central
61.	G17F2	A.b.3.3.	Comblement du trou central
62.	G18F2	A.b.3.4.	Excentration du trou central
63.	S25L6	A.b.3.4.	Déformation du trou central
64.	G19F2	A.b.3.8.	Accumulation de Pu dans le trou central
65.	G20F2	A.b.3.9.	Restructuration (sur défaut initial)
66.	G21F2	A.b.3.10.	Fusion
67.	S26L6	A.b.3.11.	Désintégration
68.	G22F2	A.b.3.12.	Migration du Pu (le long du trou central)

A.COMBUSTIBLE b. Annulaire 4. Cavité

69.	G23F2	A.b.4.3.	Trou central (élargissement progressif à travers plusieurs pastilles)

A.COMBUSTIBLE c. Jeu entre pastilles 0. A l'état initial

70.	S16L6	A.c.0.
71.	S17L6	A.c.0.

A.COMBUSTIBLE c. Jeu entre pastilles 3. Changement de forme ou de place

72.	R9L3	A.c.3.1.	Elargissement
73.	S18L6	A.c.3.1.	Elargissement

A.COMBUSTIBLE c. Jeu entre pastilles 8. Fluide primaire

74.	G11L2	A.c.8.1.	Rempli d'eau (plusieurs jeux)
75.	P22L5	A.c.8.1.	Rempli d'eau (état antérieur en P23L5)
76.	G12L2	A.c.8.2.	Vide d'eau (un seul jeu)
77.	P23L5	A.c.8.2.	Vide d'eau

A.COMBUSTIBLE d. Evidement 0. A l'état initial

78.	P24L5	A.d.0.
79.	P25L2	A.d.0.

A.COMBUSTIBLE d. Evidement 3. Changement de forme ou de place

80.	P26L2	A.d.3.3.	Disparition (état antérieur en P25L2)
81.	P27L5	A.d.3.4.	Déformation (état antérieur en P24L5)
82.	R10L3	A.d.3.4.	Déformation
83.	S19L6	A.d.3.4.	Déformation

A.COMBUSTIBLE e. Vibro-compacté 0. A l'état initial

84.	P17L4	A.e.0.	(Sol-gel)
85	P18L2	A.e.0.	(Noyaux)

A. FUEL b. Annular 3. Change of shape or location

60.	G16F2	A.b.3.1. Central void enlarged
61.	G17F2	A.b.3.3. Central void filled-up (at several places)
62.	G18F2	A.b.3.4. Central void deformed (eccentric)
63.	S25L6	A.b.3.4. Central void deformed
64.	G19F2	A.b.3.8. Pu accumulated in central void
65.	G20F2	A.b.3.9. Restructuring (at initial defect)
66.	G21F2	A.b.3.10.Melting
67.	S26L6	A.b.3.11.Disintegrated
68.	G22F2	A.b.3.12.Migration of Pu (along central void)

A. FUEL b. Annular 4. Voidage

69.	G23F2	A.b.4.3. Central void (increasing through several pellets)

A. FUEL c. Pellet-to-pellet gap 0. As fabricated

70.	S16L6	A.c.0.
71.	S17L6	A.c.0.

A. FUEL c. Pellet-to-pellet gap 3. Change of shape or location

72.	R9L3	A.c.3.1. Enlarged
73.	S18L6	A.c.3.1. Enlarged

A. FUEL c. Pellet-to-pellet gap 8. Coolant

74.	G11L2	A.c.8.1. Filled with water (several gaps)
75.	P22L5	A.c.8.1. Filled with water (original as P23L5)
76.	G12L2	A.c.8.2. Without water (one gap)
77.	P23L5	A.c.8.2. Without water

A. FUEL d. Dishing 0. As fabricated

78.	P24L5	A.d.0.
79.	P25L2	A.d.0.

A. FUEL d. Dishing 3. Change of shape or location

80.	P26L2	A.d.3.3. Disappeared (original as P25L2)
81.	P27L5	A.d.3.4. Deformed (original as P24L5)
82.	R10L3	A.d.3.4. Deformed
83.	S19L6	A.d.3.4. Deformed

A. FUEL e. Vibro-compacted 0. As fabricated

84.	P17L4	A.e.0. (sol-gel)
85.	P18L2	A.e.0. (kernels)

A.COMBUSTIBLE e. Vibro-compacté 1. Fissures

86.   P19L2   A.e.1.3.   Transversale

A.COMBUSTIBLE e. Vibro-compacté 2. Eclats

87.   P20L4   A.e.2.4.   Coin manquant

A.COMBUSTIBLE e. Vibro-compacté 4. Cavité

88.   P21L2   A.e.4.3.   Trou central au long de la colonne

A. COMBUSTIBLE f. Jeu combustible-gaine 0. A l'état initial

89.   S27F6   A.f.0.
90.   S28L6   A.f.0.
91.   R21L4   A.f.0.   (Jeu étalonné)

A.COMBUSTIBLE f. Jeu combustible-gaine 3. Changement de forme ou
de place

92.   R11L3   A.f.3.1.   Elargissement
93.   G24L2   A.f.3.1.   Elargissement
94.   H2F1   A.f.3.4.   Déformation

A.COMBUSTIBLE g. Colonne combustible 0. A l'état initial

95.   P29L2   A.g.0.   (Vibro-compacté)
96.   P30F6   A.g.0   (En pastilles)

A.COMBUSTIBLE g. Colonne combustible 3. Changement de forme
oude place

97.   H3F1   A.g.3.2.   Contraction
98.   G25L2   A.g.3.4.   Déformation
99.   P31F2   A.g.3.4.   Déformation (combustible fondu)
100.   P32F2   A.g.3.4.   Déformation (séparation de la colonne)
101.   G26F2   A.g.3.7.   Allongement (par fusion)
102.   G27F2   A.g.3.10. Fusion complète de la colonne
103.   P33F2   A.g.3.10. Fusion (et disparition partielle)
104.   P34F2   A.g.3.11. Désintégration

A.COMBUSTIBLE h. Constitution 0. A l'état initial

105.   S29L6   A.h.0.   (Avec poison Gd granulaire)

A.COMBUSTIBLE h. Constitution 3. Changement de forme ou de place

106.   G28F2   A.h.3.12. Migration du Pu (dans les fissures transversales)
107.   G29F2   A.h.3.12. Migration du Pu (le long du trou central)
108.   G30F2   A.h.3.12. Migration du Pu (le long du trou central)

A. FUEL e. Vibro-compacted 1. Cracks

86.   P19L2   A.e.1.3. Transverse

A. FUEL e. Vibro-compacted 2. Chips

87.   P20L4   A.e.2.4. Missing

A. FUEL e. Vibro-compacted 4. Voidage

88.   P21L2   A.e.4.3. Central void through column

A. FUEL f. Fuel-to-clad gap 0. As fabricated

89.   S27F6   A.f.0.
90.   S28L6   A.f.0.
91.   R21L4   A.f.0.   (calibrated gap)

A. FUEL f. Fuel-to-clad gap 3. Change of shape or location

92.   R11L3   A.f.3.1. Enlarged
93.   G24L2   A.f.3.1. Enlarged
94.   H2F1   A.f.3.4. Deformed

A. FUEL g. Fuel column 0. As fabricated

95.   P29L2   A.g.0.   (vibro-compacted)
96.   P30F6   A.g.0.   (pelletized)

A. FUEL g. Fuel column 3. Change of shape or location

97.   H3F1   A.g.3.2. Contracted
98.   G25L2   A.g.3.4. Deformed
99.   P31F2   A.g.3.4. Deformed (fuel melted)
100.   P32F2   A.g.3.4. Deformed (stack separation)
101.   G26F2   A.g.3.7. Extended (by melting)
102.   G27F2   A.g.3.10.Melted (whole column)
103.   P33F2   A.g.3.10.Melted (and partly disappeared)
104.   P34F2   A.g.3.11.Disintegrated

A. FUEL h. Composition 0. As fabricated

105.   S29L6   A.h.0.   (with granular Gd poison)

A. FUEL h. Composition 3. Change of shape or location

106.   G28F2   A.h.3.12.Migration of Pu (in transverse cracks)
107.   G29F2   A.h.3.12.Migration of Pu (along central void)
108.   G30F2   A.h.3.12.Migration of Pu (along central void)

A.COMBUSTIBLE h. Constitution 5. Inclusions

109.	S30L6	A.h.5.1.	De plutonium
110.	S31L6	A.h.5.2.	De poison

B.GAINE 0. A l'état initial

111.	P35L4	B.0.	(Crayon combustible REP)

B.GAINE 3. Changement de forme ou de place

112.	R12L3	B.3.4.	Déformation
113.	R13L3	B.3.5.	Rupture
114.	G31L2	B.3.5.	Rupture
115.	G32F2	B.3.5.	Rupture
116.	G33L2	B.3.5.	Rupture
117.	G34L2	B.3.7.	Distension

B.GAINE 6. Corrosion

118.	G35L2	B.6.1.	Hydrures (au niveau des jeux entre pastilles)
119.	S32L6	B.6.1.	Hydrures
120.	S33L6	B.6.1.	Hydrures

C.VOLUME LIBRE a. Ressort 0. A l'état initial

121.	P36L5	C.a.0.	(Crayon expérimental REB)
122.	P37L2	C.a.0.	(Crayon expérimental REP)

C.VOLUME LIBRE a. Ressort 3. Changement de forme ou de place

123.	R14L3	C.a.3.2.	Compression
124.	S34L6	C.a.3.2.	Compression
125.	S35L6	C.a.3.4.	Déformation
126.	R15L3	C.a.3.6.	Dislocation
127.	G36F2	C.a.3.11.	Désintégration (par fusion)

C.VOLUME LIBRE a. Ressort 8. Fluide primaire

128.	G37L2	C.a.8.1.	Sur les spires du ressort
129.	P38L5	C.a.8.1.	Dans le ressort (noyé)

C.VOLUME LIBRE b. Manchon du ressort 0. A l'état initial

130.	S36L6	C.b.0.	

C.VOLUME LIBRE b. Manchon du ressort 3. Changement de forme ou de place

131.	R16L3	C.b.3.5.	Rupture

A. FUEL h. Composition 5. Inclusions

| 109. | S30L6 | A.h.5.1. | Of Pu |
| 110. | S31L6 | A.h.5.2. | Of poison |

B. CLADDING 0. As fabricated

| 111. | P35L4 | B.0. | (PWR fuel rod) |

B. CLADDING 3. Change of shape or location

112.	R12L3	B.3.4.	Deformed
113.	R13L3	B.3.5.	Broken
114.	G31L2	B.3.5.	Broken
115.	G32F2	B.3.5.	Broken
116.	G33L2	B.3.5.	Broken
117.	G34L2	B.3.7.	Extended

B. CLADDING 6. Corrosion

118.	G35L2	B.6.1.	Hydrides (at pellet-to-pellet interfaces)
119.	S32L6	B.6.1.	Hydrides
120.	S33L6	B.6.1.	Hydrides

C. PLENUM a. Spring 0. As fabricated

| 121. | P36L5 | C.a.0. | (BWR test fuel rod) |
| 122. | P37L2 | C.a.0. | (PWR test fuel rod) |

C. PLENUM a. Spring 3. Change of shape or location

123.	R14L3	C.a.3.2.	Contracted
124.	S34L6	C.a.3.2.	Contracted
125.	S35L6	C.a.3.4.	Deformed
126.	R15L3	C.a.3.6.	Dislocated
127.	G36F2	C.a.3.11.	Disintegrated (by melting)

C. PLENUM a. Spring 8. Coolant

| 128. | G37L2 | C.a.8.1. | On spring |
| 129. | P38L5 | C.a.8.1. | In spring (flooded) |

C. PLENUM b. Spring sleeve 0. As fabricated

| 130. | S36L6 | C.b.0. | |

C. PLENUM b. Spring sleeve 3. Change of shape or location

| 131. | R16L3 | C.b.3.5. | Broken |

C.VOLUME LIBRE c. Disque d'isolement 0. A l'état initial

132.	P39L4	C.c.0.	(Extrémité supérieure de combustible vibro-compacté)
133.	P40L5	C.c.0.	(Extrémité inférieure de combustible en pastilles)
134.	P41L4	C.c.0.	(Extrémité supérieure de combustible en pastilles)

C.VOLUME LIBRE c. Disque d'isolement 3. Changement de forme ou de place

135.	P42L5	C.c.3.5.	Rupture
136.	R17L3	C.c.3.6.	Dislocation

C.VOLUME LIBRE c. Disque d'isolement 6. Corrosion

137.	G38L2	C.c.6.1.	Interaction avec combustible

C.VOLUME LIBRE d. Séparateur 0. A l'état initial

138.	S37F6	C.d.0.	
139.	S38L6	C.d.0.	

C.VOLUME LIBRE d. Séparateur 8. Fluide primaire

140.	P43L2	C.d.8.1.	Dans le séparateur (partie basse)
141.	P44L2	C.d.8.2.	Absent (P43L2 avant irradiation)

C.VOLUME LIBRE e. Colonne combustible au bouchon 0. A l'état initial

142.	R22L3	C.e.0.	

C.VOLUME LIBRE e. Colonne combustible au bouchon 8. Fluide primaire

143.	G39L2	C.e.8.1.	Dans le volume libre

D.BOUCHON a. Extrémité inférieure 0. A l'état initial

144.	P45L5	D.a.0.	(Crayon expérimental REP)

D.BOUCHON a. Extrémité inférieure 2. Eclats

145.	S39L6	D.a.2.2.	Dans le bouchon

D.BOUCHON a. Extrémité inférieure 6. Corrosion

146.	R18L3	D.a.6.1.	Hydrures

D.BOUCHON b. Extrémité supérieure 0. A l'état initial

147.	P46L4	D.b.0.	(Crayon expérimental)
148.	S40F6	D.b.0.	(Avec thermocouple)

C. PLENUM c. Insulating disc 0. As fabricated

132. P39L4   C.c.0.   (top of vibro-compacted fuel)
133. P40L5   C.c.0.   (bottom of pelletized fuel)
134. P41L4   C.c.0.   (top of pelletized fuel)

C. PLENUM c. Insulating disc 3. Change of shape or location

135. P42L5   C.c.3.5.   Broken
136. R17L3   C.c.3.6.   Dislocated

C. PLENUM c. Insulating disc 6. Corrosion

137. G38L2   C.c.6.1.   Interaction with fuel

C. PLENUM d. Spacer 0. As fabricated

138. S37F6   C.d.0.
139. S38L6   C.d.0.

C. PLENUM d. Spacer 8. Coolant

140. P43L2   C.d.8.1.   In spacer (at bottom)
141. P44L2   C.d.8.2.   Absent (P43L2 - not irradiated)

C. PLENUM e. Fuel column-to-plug 0. As fabricated

142. R22L3   C.e.0.

C. PLENUM e. Fuel column-to-plug 8. Coolant

143. G39L2   C.e.8.1.   In plenum

D. PLUG a. Bottom 0. As fabricated

144. P45L5   D.a.0.   (PWR test fuel rod)

D. PLUG a. Bottom 2. Chips

145. S39L6   D.a.2.2.   In plug

D. PLUG a. Bottom 6. Corrosion

146. R18L3   D.a.6.1.   Hydrides

D. PLUG b. Top 0. As fabricated

147. P46L4   D.b.0.   (test fuel rod)
148. S40F6   D.b.0.   (with thermocouple)

27

D.BOUCHON b. Extrémité supérieure 3. Changement de forme ou de place

149. R19L3 D.b.3.5. Rupture

D.BOUCHON b. Extrémité supérieure 6. Corrosion

150. R20L3 D.b.6.1. Hydrures

E.INSTRUMENTATION a. Thermocouple 0. A l'état initial

151. S41F6 E.a.0
152. P47F6 E.a.0.

E.INSTRUMENTATION a. Thermocouple 3. Changement de forme ou de place

153. S43L6 E.a.3.10. Fusion

E. INSTRUMENTATION b. Capteur de pression 0. A l'état initial

154. P48L2 E.b.0.   (Capteur de pression à membrane)

E.INSTRUMENTATION c. Jauge de diamètre 0. A l'état initial

155. G40L2 E.c.0.   Système à jauge de contrainte (capteur)

E.INSTRUMENTATION d. Jauge de longueur 0. A l'état initial

156. P49L2 E.d.0.   (Partie centrale)

E.INSTRUMENTATION e. Autre 0. A l'état initial

157. P50L2 E.e.0.   (Jonction Zr/acier inoxydable)
158. P51L2 E.e.0.   (Soufflet de mesure de volume vacant)

D. PLUG b. Top 3. Change of shape or location

149. R19L3 D.b.3.5. Broken

D. PLUG b. Top 6. Corrosion

150. R20L3 D.b.6.1. Hydrides

E. INSTRUMENTATION a. Thermocouple 0. As fabricated

151. S41F6 E.a.0.
152. P47F6 E.a.0.

E. INSTRUMENTATION a. Thermocouple 3. Changes of shape or
location

153. S43L6 E.a.3.10.Melted

E. INSTRUMENTATION b. Pressure transducer 0. As fabricated

154. P48L2 E.b.0. (membrane pressure transducer)

E. INSTRUMENTATION c. Diameter gauge 0. As fabricated

155. G40L2 E.c.0. (strain gauge sensor)

E. INSTRUMENTATION d. Length gauge 0. As fabricated

156. P49L2 E.d.0. (only core)

E. INSTRUMENTATION e. Other 0. As fabricated

157. P50L2 E.e.0. (Zr/stainless steel joint)
158. P51L2 E.e.0. (bellows system for void volume measurement)

7. TERMINOLOGIE

Le texte de ce recueil est présenté à la fois en français et en anglais. Les termes particuliers utilisés dans le cours de l'ouvrage, de même que certains termes courants dans le domaine de la neutronographie, sont repris dans le lexique ci-après en allemand, anglais, danois, français, hollandais et italien. Les traductions des termes anglais ont été établies par les personnes ci-dessous, qui ont ainsi contribué à l'ouvrage:
- Allemand: J.F.W. Markgraf, Centre Commun de Recherche Petten;
- Danois: J. Olsen, Laboratoire National de Risø;
- Français: R. Barbalat, Centre d'Etudes Nucléaires de Saclay;
- Néerlandais: H.P. Leeflang, Fondation Hollandaise pour la Recherche sur l'Energie (ECN), Petten;
- Italien: G. Trezza, Commissariat National pour la Recherche et pour le Développement de l'Energie Nucléaire et des Energies Alternatives, Rome.

7. TERMINOLOGY

The text of this collection is produced both in English and French. Special terms used throughout the collection (as shown in table 1 and chapter 6), as well as some useful ones in the field of neutron radiography, are reproduced below in Danish, Dutch, English, French, German and Italian. The following persons have given equivalent terms to the English ones:

- Danish: J. Olsen, Risø National Laboratory,
- Dutch: H.P. Leeflang, Netherlands Energy Research Foundation, Petten,
- French: R. Barbalat, Centre d'Etudes Nucléaires de Saclay,
- German: J.F.W. Markgraf, Joint Research Centre, Petten Establishment,
- Italian: G. Trezza, Comitato Nazionale per la Ricerca e per lo Sviluppo dell'Energia Nucleare e delle Energie Alternative, Roma.

Their contribution is gratefully acknowledged.

CLASSIFICATION TERMINOLOGY
(Vertical column of table 1)

ENGLISH	DANSK	DEUTSCH
A. fuel	brændsel	Brennstoff
A.a. pellets	piller	Tabletten
A.b. annular fuel	cirkulært brændsel	ringförmiger Brennstoff
A.c. pellet-to-pellet gap	gab mellem piller	Spalt zwischen zwei Tabletten
A.d. dishing	konkavitet	kugelförmige Vertiefung
A.e. vibro-compacted	compact vibrered	vibrationsverdichtet
A.f. fuel-to-clad gap	gab mellem pille og kapslingsrør	Spalt zwischen Tabletten und Hüllrohr
A.g. fuel column	brændsels stak	Brennstoffsäule
A.h. fuel composition	brændselssammenstætning	Brennstoffzusammensetzung
B. cladding	kapslingsrør	Hüllrohr
C. plenum	frit volumen	Plenum
C.a. spring	fjeder	Feder
C.b. spring sleeve	fjeder manchet	Federführungshülse (Hülse)
C.c. insulating disc	isolerende skive	Isolationsscheibe
C.d. spacer	afstandsstykke	Abstandshalter
C.e. fuel column to plug	mellem nedepille og endeprop	Brennstoffsäule bis Endstopfen
D. plugs	endepropper	Endstopfen
D.a. bottom	nedre	unteres Ende (unten)
D.b. top	øvre	oberes Ende (oben)
E. instrumentation	instrumentering	Instrumentation
E.a. thermocouple	termoelement	Thermoelement
E.b. pressure transducer	tryktransducer	Druckaufnehmer
E.c. diameter gauge	diametermåler	Durchmessermessgerät
E.d. length gauge	længdemåler	Längenmessgerät
E.e. other	andre	andere

LEXIQUE DE LA CLASSIFICATION

(Entrée par lignes de tableau 1)

FRANCAIS	ITALIANO	NEDERLANDS
combustible	combustibile	splijtstof
pastilles	pasticche	tabletten
combustible annulaire	combustibile anulare	ringvormige splijtstof
jeu entre pastilles	gioco tra pasticche	spleet tussen tabletten
évidement	svuotamento	schotelvormig
vibro-compacté	vibro-compatto	tril-verdicht
jeu combustible-gaine	gioco combustibile-guaina	spleet tussen splijtstof en omhulling
colonne combustible	colonna di combustibile	splijtstof kolom
constitution du combustible	composizione del combustibile	splijtstof samenstelling
gaine	guaina	omhulling
volume libre	volume libero	plenum
ressort	molla	veer
manchon du ressort	manicotto per molla	veerbus
disque d'isolement	rondella isolante	isolerende schijf
séparateur	spaziatore	vulstuk
colonne combustible au bouchon	colonna di combustibile al tappo	top splijtstofkolom tot eindplug
bouchons	tappi	eindpluggen
extrémité inférieure	estremitá inferiore	onderkant
extrémité supérieure	estremitá superiore	bovenkant
instrumentation	strumentazione	instrumentatie
thermocouple	termocoppia	thermokoppel
capteur de pression	transduttore di pressione	druk opnemer
jauge du diamétre	comparatore per diametri	diameter meter
jauge de longeur	comparatore per lunghezza	lengte meter
autre	altro	andere

CLASSIFICATION TERMINOLOGY
(Horizontal column of table 1)

ENGLISH	DANSK	DEUTSCH
0. as fabricated	som fremstillet	wie hergestellt
1. cracks	revner	Risse
1.1. random	tilfældig	zufällig, willkürlich
1.2. longitudinal	langsgående	längs
1.3. transverse	tværgående	quer
1.4. annular	ringformet	ringförmig
1.5. stratified	lagdelt	geschichtet
2. chips	spåner	Abplatzungen
2.1. corner	hjørne	Ecke
2.2. other	andre	andere
2.3. in pellet-to-pellet gap	i gabet mellem piller	im Spalt zwischen zwei Tabletten
2.4. missing	manglende	fehlend
3. change of shape or location	ændring af form eller placering	Form- oder Lage-änderung
3.1. enlarged or swollen	forstørret eller svulmet	vergrössert oder geschwollen
3.2. contracted	sammentrukket	zusammengedrückt
3.3. filled-up or closed	fyldt op eller lukket	aufgefüllt oder geschlossen
3.4. deformed	deformeret	verformt
3.5. broken	brækket	zerbrochen
3.6. dislocated	skubbet	verschoben (versetzt)
3.7. extended	udvidet	ausgedehnt
3.8. accumulated	akkumuleret	versammelt (konzentriert)
3.9. restructured	genopbygget	umstrukturiert (restrukturiert)
3.10. melted	smeltet	geschmolzen
3.11. disintegrated	henfaldet	zerstört (desintegriert)
3.12. migrated	diffunderet	gewandert (diffundiert)

LEXIQUE DE LA CLASSIFICATION

(Entrée par colonnes de tableau 1)

FRANCAIS	ITALIANO	NEDERLANDS
à l'état initial	come fabbricato	zoals gefabriceerd
fissures	fratture	scheuren
aléatoire	casuale	willekeurig
longitudinal	longitudinale	langsrichting
transversal	transversale	dwarsrichting
circonférentiel	anulare (circolare)	ringvormig
stratifié	stratificato	gelaagd
éclats	frammenti	spaanders
angle	angolo	hoek
autre	altro	ander
dans jeu entre pastilles	nel gioco tra pasticche	spleet ruimte tussen tabletten
manquant	mancante	ontbrekende
changement de forme	cambiamento di forma	verandering van vorm
ou de place	o posizione	of plaats
élargi ou gonflé	ingrossato o rigonfio	vergroot of gezwollen
tassé ou réduit	ridotto	samengedrukt
comblé ou refermé	riempito o richiuso	opgevuld of gesloten
déformé	deformato	gedeformeerd
rompu	rotto	gebroken
déplacé ou disloqué	dislocato	verplaatst
allongé ou distendue	esteso	verlengd (vergroot)
accumulé	accumulato	verzameld
restructuré	ristrutturato	gerestructureerd
fondu	fuso	gesmolten
désintegré	disintegrato	gedesintegreerd
ayant migré	migrato	gemigreerd

ENGLISH	DANSK	DEUTSCH
4. voidage	tomrum	Leerstellen (Lücken)
4.1. in one pellet	i en pille	in einer Tablette
4.2. through several pellets	gennem flere piller	durch mehrere Tabletten
4.3. through whole fuel column	gennem hele brændsels stakken	durch gesamte Brennstoff-säule
5. inclusions	indkapslinger	Einschlüsse
5.1. plutonium	plutonium	Plutonium
5.2. poison	gift	Gift
5.3. other	andre	andere
6. corrosion	korrosion	Korrosion
6.1. hydrides	hydrider	Hydride
6.2. oxides	oxider	Oxyde
6.3. other	andre	andere
7. neutron properties	neutron egenskaber	Eigenschaften der Neutronen
7.1. enrichment	berigning	Anreicherung
7.2. burn-up	udbrænding	Abbrand
8. coolant	kølemiddel	Kühlmittel
8.1. present	nærværende, tilstede	vorhanden
8.2. absent	fraværende, ikke tilstede	nicht vorhanden

FRANCAIS	ITALIANO	NEDERLANDS
cavité	svuotamento	leegte
dans une pastille	in una pasticcha	in één tablet
à travers plusieurs pastilles	attraverso varie pasticche	door verschillende tabletten
à travers toute la colonne	attraverso tutta la colonna	door de hele splijtstof-
combustible	combustible	kolom
inclusions	inclusioni	insluitingen
plutonium	plutonio	plutonium
poison	veleno	vergif
autre	altro	andere
corrosion	corrosione	corrosie
hydrures	idruri	hydride
oxydes	ossidi	oxyde
autre	altro	andere
qualités neutroniques	proprietá neutroniche	eigenschappen van de neutronen
enrichissement	arricchimento	verrijking
consommation,	tasso di combustione,	opbrand
taux de combustion		
fluide primaire	refrigerante	koelmiddel
présent	presente	aanwezig
absent	assente	afwezig

ADDITONAL TERMINOLOGY

ENGLISH	DANSK	DEUTSCH
assessment	vurdering	Auswertung (Beurteilung)
beam purity indicator	stralings indikator	Strahlreinheitsindikator
calibrated standard	kalibreret standard	geeichter (kalibrierter)
defect	defekt	Standard-Fehler
calibrated	kalibreret	geeicht (kalibriert)
cladding tube	kapslingsrør	Hüllrohr
calibration fuel pin	kalibreringsstav	Eichbrennstab
defects revealed by NR	defekt afsløret ved NR	durch NR festgestellte Fehler (Defekte)
double beam scanning microdensitometer	dobbelstråle skanderings-mikrodensitometer	Mikrodensitometer mit Doppelstrahlabtaster
fuel pellet	brændselspille	Brennstofftablette
fuel pin	brændselsstav	Brennstab
gap gauge	gabmål	Spaltlehre (Eichspalt)
hole gauge	hulmål	Eichbohrung
image analyzer	billed analysator	Bildanalysator
image enhancement	billed fremhævning	Bildverstärkung (Aufnahmeverstärkung)
image quality indicator	billed kvalitets indikator	Bildqualitätsindikator
light table micrometer	lysbords mikrometer	Mikrometer mit Lichtkasten
micrometer screw	mikrometerskrue	Mikrometerschraube
neutron beam	neutronstråle	Neutronenstrahl
optical projector	optisk projektor	optischer Projektor (Profilprojektor)
photographic enlargement	fotografisk forstørrelse	photographische Vergrösserung
photographic sharpening	fotografisk skarpheds-forbedring	photographische Schärfe
post-irradiation	efter bestråling	nach der Bestrahlung
pre-irradiation	før bestråling	vor der Bestrahlung
print-out device	udløser enhed	Drucker
radiation beam	stråle	Strahlenbündel

LEXIQUE COMPLEMENTAIRE

FRANCAIS	ITALIANO	NEDERLANDS
évaluation	valutazione	beoordeling
indicateur de pureté de faisceau	indicatore di purezza del fascio (raggio)	bundel zuiverheids-indikator
tube-gaine (tube de gainage)	tubo di guaina	bekledingsbuis, huls
calibré ou etalloné	calibrato	gecalibreerd
défaut standard calibré	difetti standard calibrati	gecalibreerde standaard defekt
crayon combustible calibré	barretta di combustibile calibrata	calibratie splijtstofpen
défauts révélés par la NR	difetti rilevati per mezzo dalla NR	defecten aangetoond met NR
microdensitomètre à double balayage	microdensitometro a doppia scansione (a doppio raggio)	dubbele bundel (aftast microdensitometer)
pastille combustible	pasticca di combustibile	splijtstoftablet
crayon (aiguille) combustible	barretta de combustibile	splijtstofstaaf
jauge d'épaisseur (jauge à brèche)	caiibro per spessori	spleet kaliber
jauge à trou	calibro per fori	pen kaliber
analyseur d'image	analizzatore d'immagine	beeld analysator
accroissement d'image (intensification ou amélioration d'image)	intensificazione d'immagine	beeldversterking
indicateur de qualité d'image	indicatore della qualitá dell'immagine	beeldkwaliteit indikator
micromètre à table lumineuse	micrometro a tavola luminosa	lichttafel micrometer
vis micrometrique	vite micrometrica	micrometer schroef
faisceau de neutrons	fascio di neutroni	neutronenbundel
projecteur optique	proiettore ottico	optische projector
agrandissement photographique	ingrandimento fotografica	fotografische vergroting
aiguisement photographique	risoluzione fotografico	fotografische scherpte
après irradiation	post-irraggiamento	na de bestraling
avant irradiation	pre-irraggiamento	voor de bestraling
appareil de repro-duction (duplication)	duplicatore	afdruk apparaat
faisceau de radiations	fascio di radiazioni	stralingsbundel

scanning	skandering	abtasten
scanning slit	lysåbning	Abtasterschlitz
sensitivity indicator	følsomheds indikator	Empfindlichkeitsindikator
shim	mellemlægsplade	Einstellplatte (Justierplatte)
silver halide film	sølv-halogen film	Silberhalogenid-Film

travelling miocrodensito-meter	bevægeligt mikrodensito-meter	Mikrodensitometer mit Verstelltisch
travelling microscope	bevægeligt mikroskop	Mikroskop mit Verstelltisch

balayage	scansione	aftasten
fente de balayage	fenditura di scansione (fessura)	spleet-aftaster
indicateur de sensibilité	indicatore di sensibilita	gevoeligheids indikator
cale	spessore	afstelplaatje
film argentique (film á halogénure d'argent)	film ad alogenuro d'argento	zilverhalogeen film

microdensitomètre à balayage	microdensitometro a scansione	aftast microdensitometer
microscrope à balayage	microscopio a scansione	aftast microscoop

8. INSTALLATIONS DE NEUTRONOGRAPHIE A L'INTERIEUR DE LA COMMUNAUTE EUROPEENNE

Les tableaux suivants présentent une synthèse des prinicpales charactéristiques et des adresses des installations neutronographiques permettant l'examen des combustibles nucléaires dans la Communauté Européenne.
On trouvera plus de détails techniques dans la référence [2].

8. NEUTRON RADIOGRAPHY INSTALLATIONS IN THE EUROPEAN COMMUNITY

A survey on the main technical data and addresses of the neutron radiography installations in the European Community applicable for examination of nuclear reactor fuel is given in the tables below.
More technical details can be found in reference [2].

TABLEAU 2 PRINCIPALES CARACTERISTIQUES DES INSTALLATIONS DE NEUTRONOGRAPHIE DANS LA COMMUNAUTE EUROPEENNE POUR L'EXAMEN DE COMBUSTIBLE NUCLEAIRE

Etat membre de la C.E.	Lieu	Implantation	Dispositif	Dimensions du faisceau dans le plan de l'objet (largeur x hauteur) mm.	Longueur max. du crayon combustible manipulable mm.	REP/REB non irradié	REP/REB irradié	RNR non irradié	RNR irradié	Remarques
Belgique	Mol	BR 1	à sec	300 x 300	3000	oui	non	oui	non	
		BR 2	immergé	100 x 600	3000	oui	oui	oui	oui	
Danemark	Risø	DR 1	à sec	100 x 100 (2 fois)	4500	oui	oui	oui	oui	Double faisceau de neutrons.
République Fédérale d'Allemagne Occidentale	Geesthacht	FRG 2	immergé	100 x 400	1700	oui	oui	oui	oui	Installé en cellule chaude, principale application: assemblages combustibles et barres de contrôle.
		source Sb-Be de 1 kCi	à sec	400 x 400	3000	oui	oui	oui	oui	
France	Cadarache	LDAC	à sec	500 x 100	1500	oui	oui	oui	oui	Risque de contamination.
		SCARABEE	à sec	180 x 240	1500	oui	oui	oui	oui	En préparation, pour assemblages combustible ø 120 x 1500.
	Grenoble	MELUSINE	à sec	240 x 180 ø max. 400	4000	oui	non	oui	non	Utilisation habituelle pour l'industrie et pour contrôle quantitatif par transmission d'un faisceau de neutrons froids.

Site	Réacteur	Type	Dimensions	(mm)	Utilisation habituelle pour le suivi de combustibles en dispositifs d'irradiation, avec possibilité de mesures dimensionnelles.	Caméra automatique pour séries de neutronographies (35 clichés/4000 mm) sur 5 aiguilles combustibles placées côte à côte.	Examen des dispositifs irradiés à OSIRIS (installation pouvant remplacer celle d'OSIRIS).	Utilisation industrielle à balayage pour matériaux non radioactifs.	Contrôle des combustibles en dispositif d'irradiation.		
Saclay	SILOE	immergé	120 x 400	1800	oui	oui	oui	non	oui		
	ISIS	à sec	100 x 150	4000	oui	non	oui	oui	oui	oui	oui
	ISIS	immergé	150 x 600	1800	oui	oui	oui	non	oui		
	ORPHEE	à sec	25 x 150	4000	oui	non	oui	oui	oui	oui	oui
	OSIRIS	immergé	100 x 650	1800	oui	oui	oui	non	oui		
Valduc	MIRENE	à sec	180 x 240 (faisceau tangentiel) 300 x 300 (faisceau axial)	2000 2000							

Etat membre de la C.E.	Lieu	Implantation	Dispositif	Dimensions du faisceau dans le plan de l'objet (largeur x hauteur) mm.	Longueur max. du crayon combustible manipulable mm.	REP non irradié	REP irradié	RNR non irradié	RNR irradié	Remarques
Italie	Casaccia	TRIGA-RC1	à sec	120 ø	520		oui			Non opérationnel actuellement.
Pays-Bas	Petten	HFR(PSF)	immergé	80 x 600	1560	oui	oui	oui	oui	
		HFR(HB8)	à sec	160 x 120	2500	oui	oui	oui	oui	Caméra automatique pour séries de neutronographies sur 4 aiguilles combustibles placées côte à côte.
		LFR	à sec	250 ø	7500	oui	non	oui	non	
Royaume-Uni	Harwell	DIDO	à sec	216 x 216	1750	oui	oui	oui	oui	

Table 2. MAIN TECHNICAL DATA OF NEUTRON RADIOGRAPHY INSTALLATIONS IN THE EUROPEAN COMMUNITY FOR EXAMINATION OF NUCLEAR FUEL

EC-member state	Site	Facility	Camera type	Beam dimensions at object plane (width x height) mm	Max. fuel rod length to be handled mm	Facility applicable for examination of fuel rods from				Remarks
						LWR		FBR		
						un-irra-diated	irra-diated	un-irra-diated	irra-diated	
Belgium	Mol	BR 1	dry	300 x 300	3000	yes	no	yes	no	
		BR 2	pool	100 x 600	3000	yes	yes	yes	yes	
Denmark	Risø	DR 1	dry	100 x 100(twice)	4500	yes	yes	yes	yes	Double neutron beam
Federal Republic of Germany	Geest-hacht	FRG 2	pool	100 x 400	1700	yes	yes	yes	yes	
		1 kCi Sb-Be neutron source	dry	400 x 400	3000	yes	yes	yes	yes	Installed in hot cell, main application on fuel bundles and control rods
France	Cada-rache	LDAC	dry	500 x 100	1500	yes	yes	yes	yes	risk of cross-contamination
		SCARABEE	dry	180 x 240	1500	yes	yes	yes	yes	In preparation, for fuel bundles 120 ø x 1500
	Grenoble	MELUSINE	dry	240 x 180 max. 400 ø	4000	yes	no	yes	no	Current use for industrial applications and quantitative testing by cold neutron transmitter beam

Country	Location	Reactor	Type	Size (mm)	Number					Remarks
	Saclay	SILOE	pool	120 x 400	1800	yes	yes	yes	yes	For inspection of fuel rods in irradiation rigs, possibility for dimensional measurements
		ISIS	dry	100 x 150	4000	no	yes	no	yes	Automatic camera for sequential neutron radiography (35 images/4000 mm), on 5 fuel rods placed side by side
		ISIS	pool	150 x 600	1800	yes	yes	yes	yes	Examination of irradiation devices from OSIRIS (back-up of ORISIS facility) Industrial applications.
		ORPHEE	dry	25 x 150	4000	yes	no	yes	no	Non radioactive components on a production line
		OSIRIS	pool	100 x 650	1800	yes	yes	yes	yes	For inspection of fuel rods in irradiation devices
	Valduc	MIRENE	dry	180 x 240 (tangential beam)	2000	yes		yes		
				300 x 300 (axial beam)	2000	yes		yes	yes	
Italy	Casaccia	TRIGA-RC 1	dry	120 ø	520		yes			At present not operational
The Nether-lands	Petten	HFR(PSF)	pool	80 x 600	1560	yes	yes	yes	yes	
		HFR(HB8)	dry	160 x 120	2500	yes	yes	yes	yes	Automatic camera for sequential neutron-radiography on 4 fuel rods placed side by side
		LFR	dry	250 ø	7500	yes	no	yes	no	
United Kingdom	Harwell	DIDO	dry	216 x 216	1750	yes	yes	yes	yes	

TABLEAU 3. ADRESSES DES INSTALLATIONS DE NEUTRONOGRAPHIE DANS LA COMMUNAUTE EUROPEENNE

Etat membre de la C.E.	Lieu	Implan-tation	Adresse	Téléphone (prefixe-local)	Télex Cable
Belgique	Mol	BR 1	CEN/SCK Reactor Physics Department Boeretang 200 B 2400 MOL	014-311801	31922 atomol
		BR 2	CEN/SCK BR2 Boeretang 200 B 2400 MOL	014-311801	31922b atomol
Danemark	Risø	DR 1	Risø National Laboratory Postbox 49 DK 4000 ROSKILDE	02-371212	43116 dk risatom
Republique Federale d'Allemagne Occidentale	Geest-hacht	FRG 2 1 kC neutron source	GKSS Institut für Physik Postfach 1160 D 2054 GEESTHACHT-Tesperhude	041-521212	18712 gkss g
France	Cada-rache	LDAC	CEN Cadarache D D E C / S L H A / L D A B.P. No. 1 F 13115 St.PAUL-LEZ-DURANCE	42-257193	440678 cen/ca
		SCARABEE	CEN Cadarache D E R S B.P. No. 1 F 13115 St.PAUL-LEZ-DURANCE	42-25-3223	440678 cen/ca

Grenoble	MELUSINE, SILOE	CEN Grenoble Service des Piles SEREG B.P. No. 85X F 38041 GRENOBLE Cédex	76-974111 (ext.4280)	320323 f energ
Saclay	ISIS OSIRIS	CEN Saclay Services des Piles F 91191 GIF-SUR-YVETTE Cédex	6-9082521	690641 f energ
	ORPHEE	CEN Saclay Service des Piles Neutronographie Industrielle F 91191 GIF-SUR-YVETTE Cédex	6-9083881	690641 f energ
Valduc	MIRENE	CEN Valduc SEESNC B.P. No. 21 F 21120 IS-SUR-TILLE	80-351305	270 746 diram paris valduc service srsc
Italie / Casaccia	TRIGA-RC 1	E.N.E.A.-CSN della Casaccia B.P. 2400 I 00100 ROMA	69481	613296 eneaca i
Pays-Bas / Petten	HFR	Centre Commun de Recherche de la CE HFR Division Postbox 2 NL 1755 ZG PETTEN or	02246-5656	57211 Bas reacp nl
	HFR LFR	ECN-Stichting Energieonderzoek Centrum Nederland Reactor Afdeling Postbox 1 NL 1755 ZG PETTEN	02246-4949	57211 reacp nl
Royaume-Uni / Harwell	DIDO	AERE Harwell Research Reactors Division OXFORDSHIRE OX11 ORA England	0235-24141 (ext. 5064)	83135 uk atomha g

Table 3 ADDRESSES OF NEUTRON RADIOGRAPHY INSTALLATIONS IN THE EUROPEAN COMMUNITY

EC-member state	Site	Facility	Address	Phone (area-local)	Telex Cable
Belgium	Mol	BR 1	CEN/SCK Reactor Physics Department Boeretang 200 B 2400 MOL	014 - 311801	31922 atomol
		BR 2	CEN/SCK BR2 Boeretang 200 B 2400 MOL	014 - 311801	31922b atomol
Denmark	Risø	DR 1	Risø National Laboratory Postbox 49 DK 4000 ROSKILDE	45-2-371212	43116 dk risatom
Federal Republic of Germany	Geest-hacht	FRG 2 1 kCi neutron source	GKSS Institut für Physik Postfach 1160 D 2054 GEESTHACHT-Tesperhude	041-52121	218712 gkss g
France	Cada-rache	LDAC	CEN Cadarache D D E C /S L H A / L D A B. P. No. 1 F 13115 St. PAUL-LEZ-DURANCE	42-257193	440678 cen/ca
		SCARABEE	CEN Cadarache D.E.R.S. B.P. No. 1 F 13115 St. PAUL-LEZ-DURANCE	42-253223	440678 cen/ca

Country	Location	Reactor	Address	Phone	Telex
	Grenoble	MELUSINE, SILOE	CEN Grenoble Service des Piles SEREG B.P. No. 85X F 38041 GRENOBLE Cédex	76-974111 (ext. 4280)	320323 f energ
	Saclay	ISIS OSIRIS	CEN Saclay Services des Piles F 91191 GIF-SUR-YVETTE Cédex	6-9082521	690641 f energ
		ORPHEE	CEN Saclay Service de Piles Neutronographie Industrielle F 91191 GIF-SUR-YVETTE Cédex	6-9083881	690641 f energ
	Valduc	MIRENE	CEN Valduc SEESNC B.P. No. 21 F 21120 IS-SUR-TILLE	80-351305	270 746 diram paris valduc service srsc
Italy	Casaccia	TRIGA-RC 1	E.N.E.A.-CSN della Casaccia B.P. 2400 I 00100 ROMA	69481	613296 eneaca i
The Netherlands	Petten	HFR	Joint Research Centre of CEC HFR Division Postbox 2 NL 1755 ZG PETTEN or	02246-5656	57211 reacp nl
		HFR LFR	ECN-Stichting Energieonderzoek Centrum Nederland Reactor Afdeling Postbox 1 NL 1755 ZG PETTEN	02246-4949	57211 reacp nl
United Kingdom	Harwell	DIDO	AERE Harwell Research Reactors Division OXFORDSHIRE OX11 ORA England	0235-24141 (ext. 5064)	83135 uk atomha g

9. REFERENCES

[1] Domanus, J.C.,
Neutron radiographic findings in light water reactor fuel, Risø National Laboratory,
Metallurgy Department,June 1979.

[2] von der Hardt, P., Röttger, H. editors,
Neutron Radiography Handbook, D. Reidel, Publishing Company, 1981, EUR 7622e,
ISBN 90-277-1378-2.

[3] Barton, J.P., von der Hardt, P., editors,
Neutron Radiography, Proceedings of the First World Conference in San Diego,
California, U.S.A., Dec. 1981. D. Reidel Publishing Company, 1982, EUR 8296 EN,
ISBN 90-277-1528-9.

10. **COLLECTION DES NEUTRONOGRAMMES SUR PAPIER PHOTOGRAPHIQUE (ECHELLE 2:1) ET FILM (ECHELLE 1:1)**

10. **REFERENCE NEUTRON RADIOGRAPHS ON PHOTOGRAPHIC PAPER (SCALE 2:1) AND FILM (SCALE 1:1)**

1. P1L4

2. P2F6

3. H1L1

A.FUEL a.Pellets 0.As fabricated

P1L4

A.a.0.

P2F6

A.a.0.

A.FUEL a.Pellets 1.Cracks

H1L1

A.a.1.1. Random

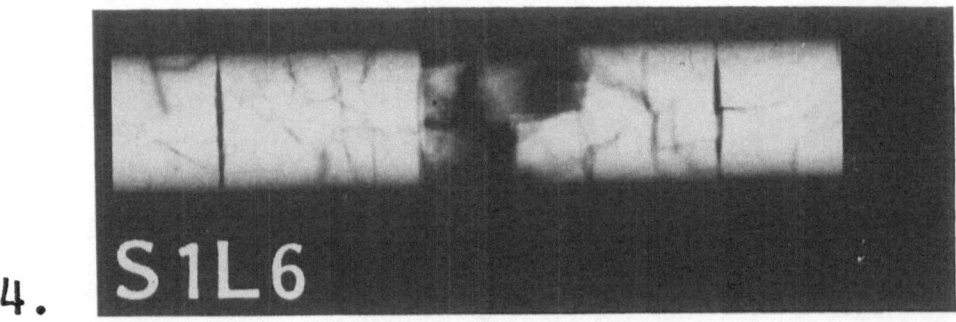

4.

S 1 L 6

5.

G 1 F 2

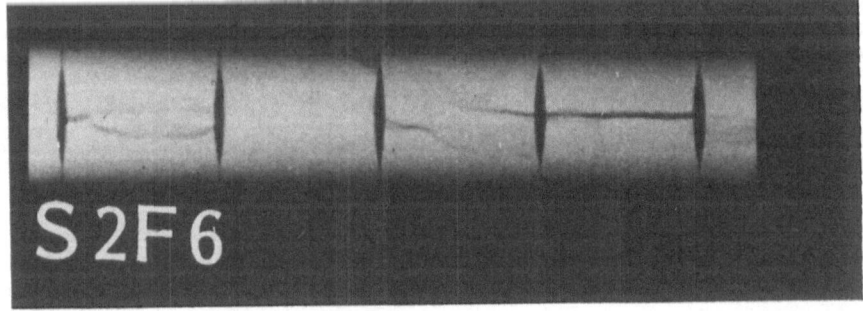

6.

S 2 F 6

A.FUEL a.Pellets 1.Cracks

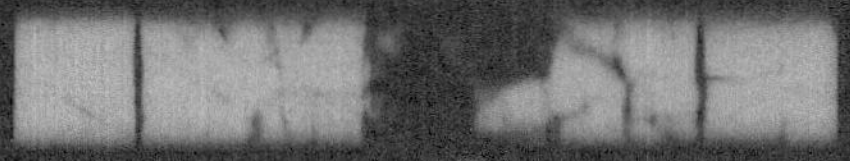

S 1 L 6

A.a.1.1. Random

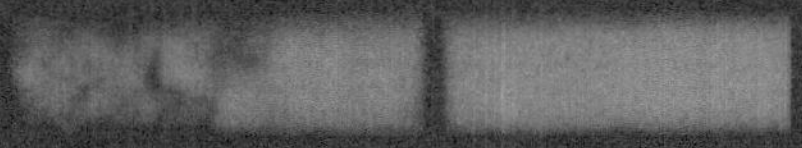

G 1 F 2

A.a.1.1. Random

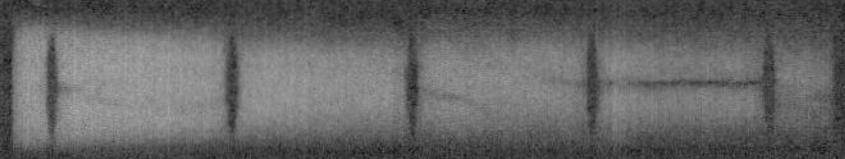

S 2 F 6

A.a.1.2. Longitudinal

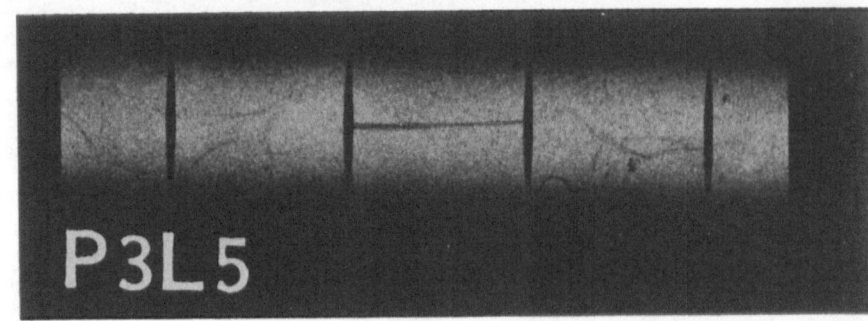

P3L5

7.

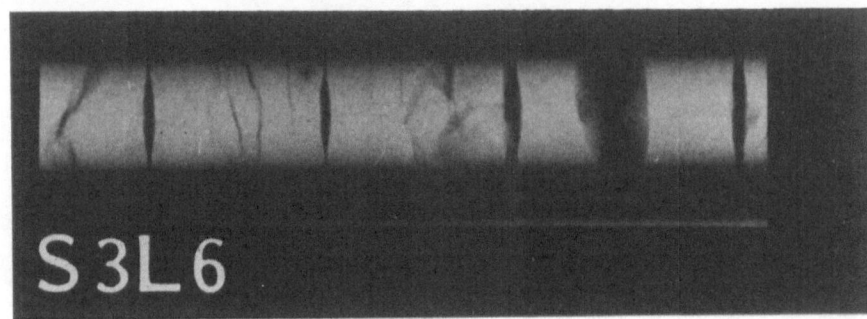

S3L6

8.

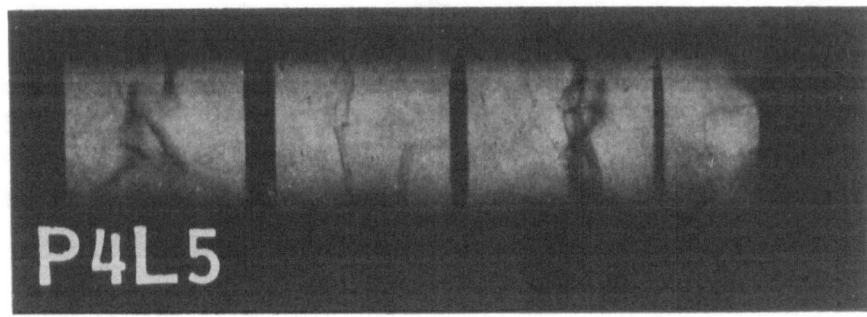

P4L5

9.

A.FUEL a.Pellets 1.Cracks

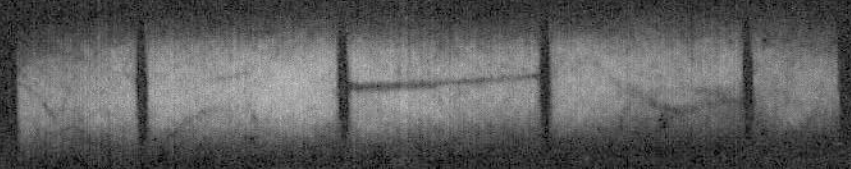

P3L5

A.a.1.2. Longitudinal

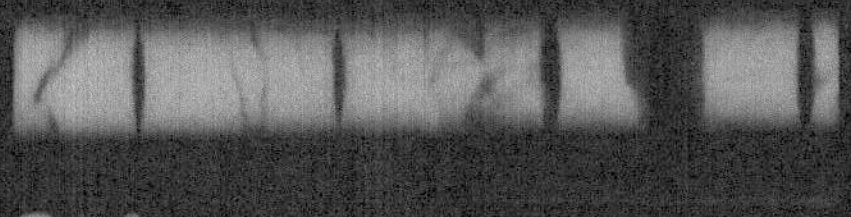

S3L6

A.a.1.3. Transverse

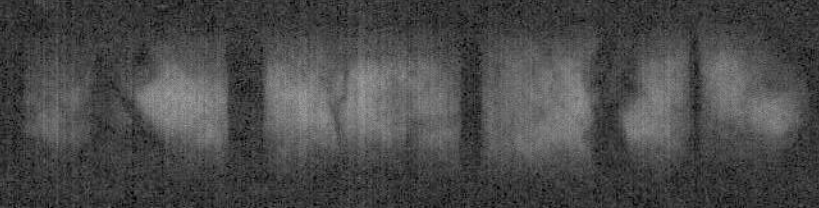

P4L5

A.a.1.3. Transverse

10.

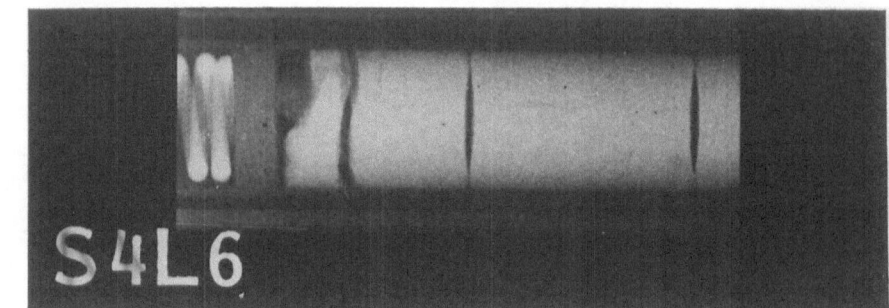

11.

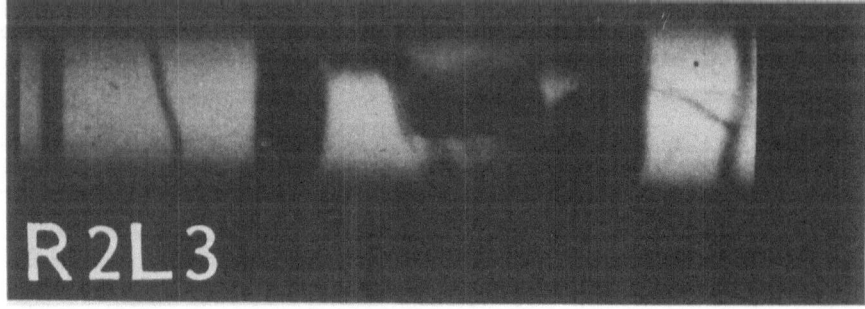

12.

A.FUEL a.Pellets 1.Cracks

R1L3

A.a.1.4. Annular

A.FUEL a.Pellets 2.Chips

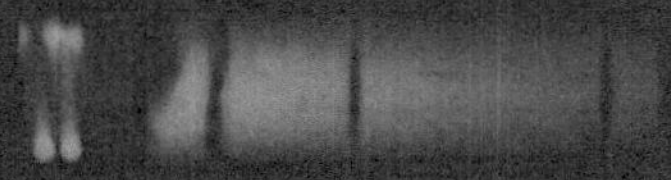

S4L6

A.a.2.1. Corner

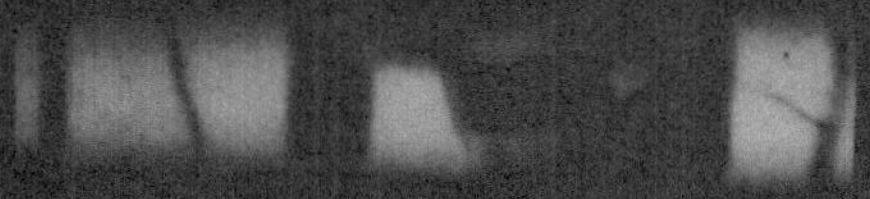

R2L3

A.a.2.2. Other

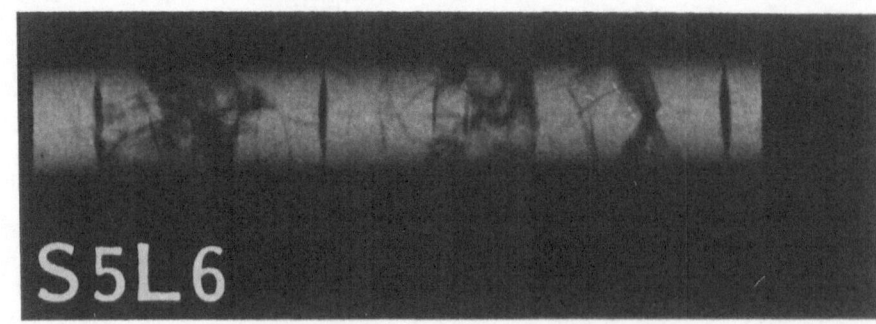

13.

14.

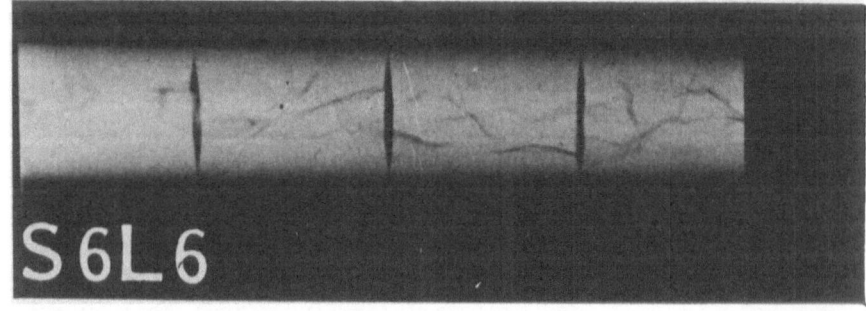

15.

A.FUEL a.Pellets 2.Chips

S 5 L 6

A.a.2.2. Other

R 3 L 3

A.a.2.3. In pellet-to-pellet gap

S 6 L 6

A.a.2.3. In pellet-to-pellet gap

16.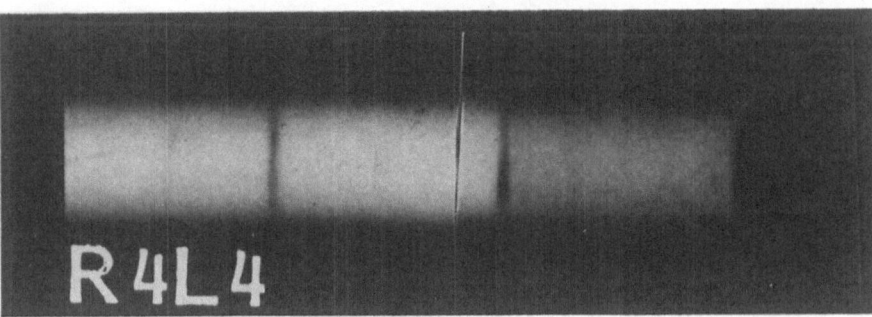

P5L5

17.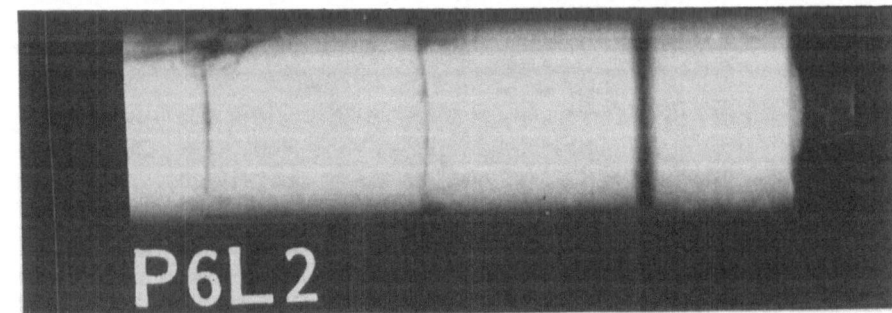

R4L4

18.

P6L2

A.FUEL a.Pellets 2.Chips

P5L5

A.a.2.3. In pellet-to-pellet gap

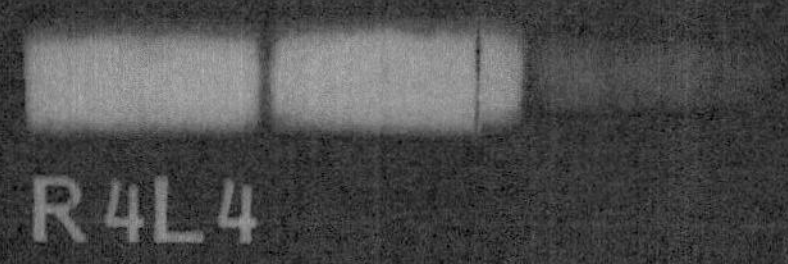

R4L4

A.a.2.4. Missing

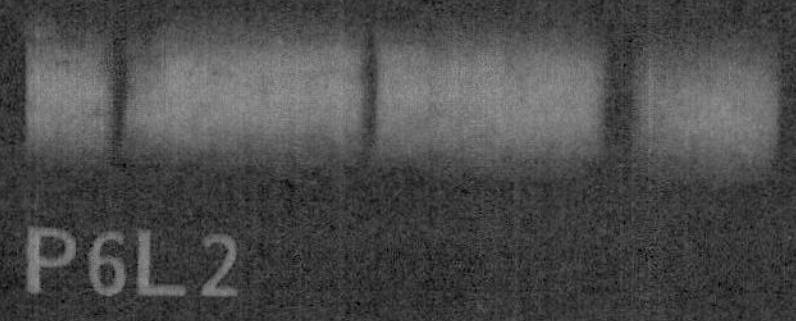

P6L2

A.a.2.4. Missing (corner)

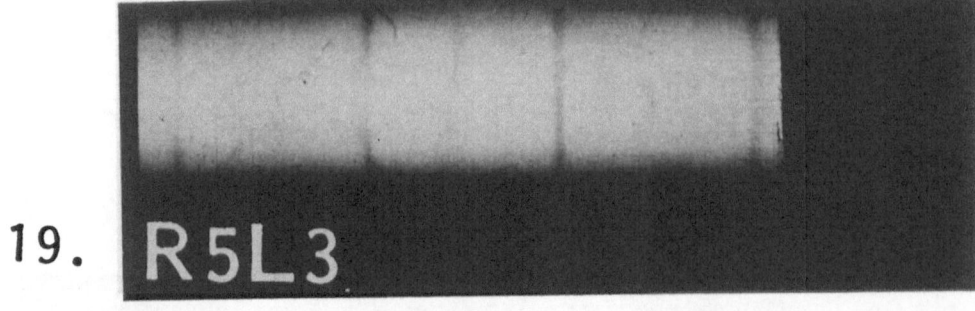

19. R5L3

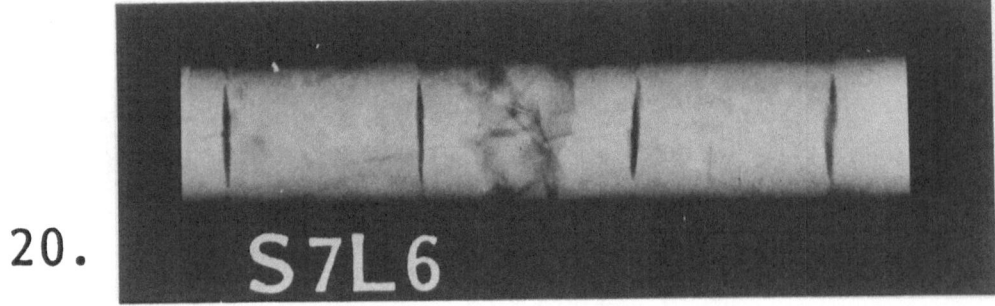

20. S7L6

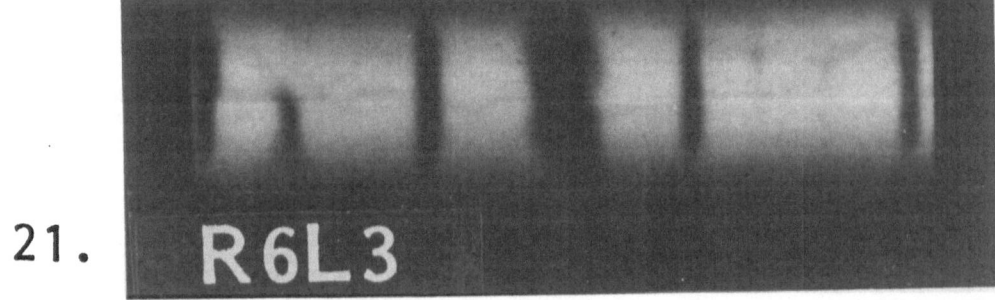

21. R6L3

A.FUEL a.Pellets 3.Change of shape or location

R5L3

A.a.3.1. Enlarged or swollen

S7L6

A.a.3.1. Enlarged or swollen

R6L3

A.a.3.5. Broken

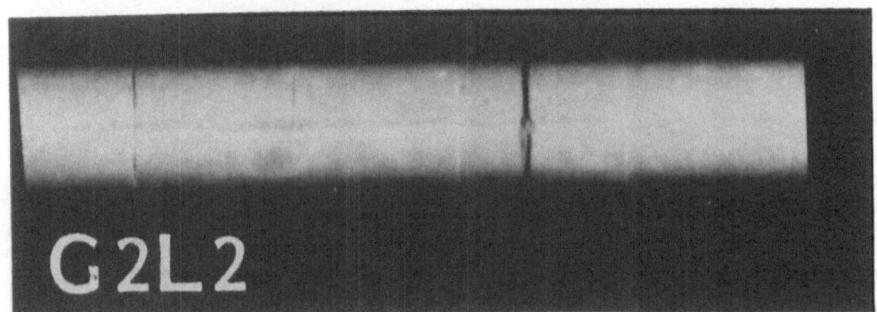

22.

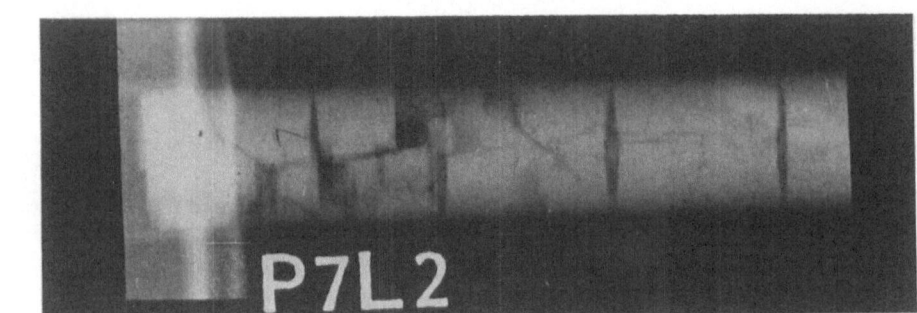

23.

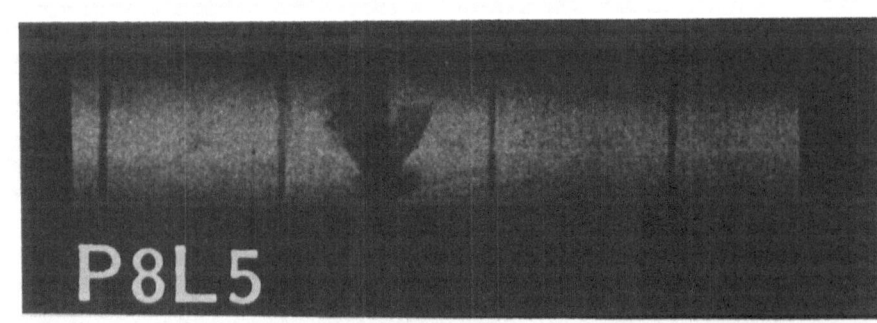

24.

A.FUEL a.Pellets 3.Change of shape or location

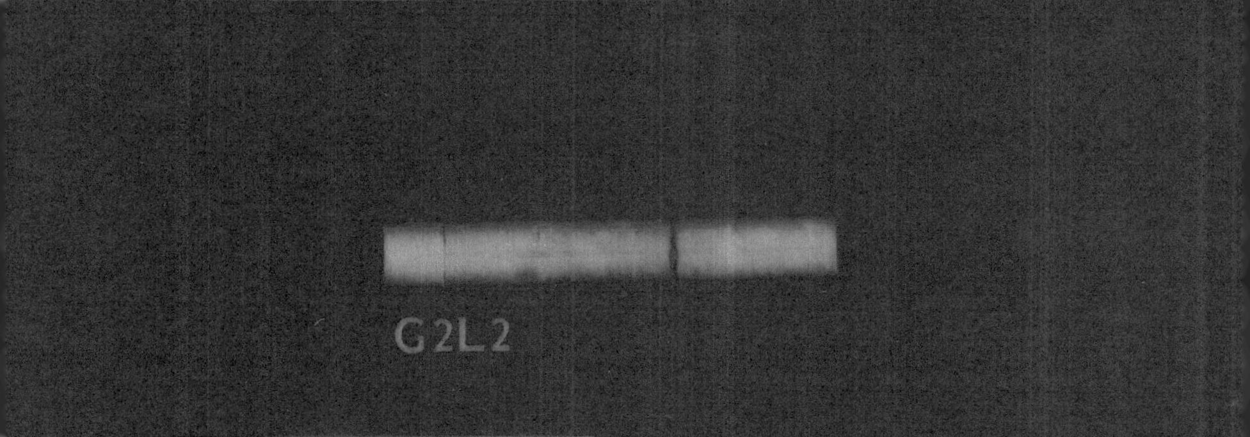

G2L2

A.a.3.5. Broken (transversly)

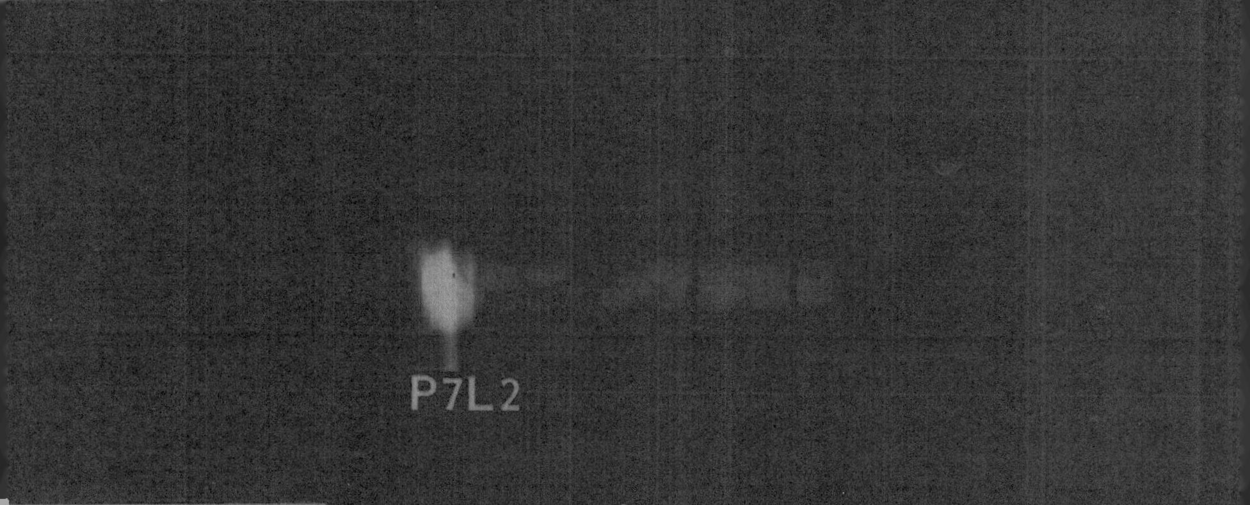

P7L2

A.a.3.5. Broken

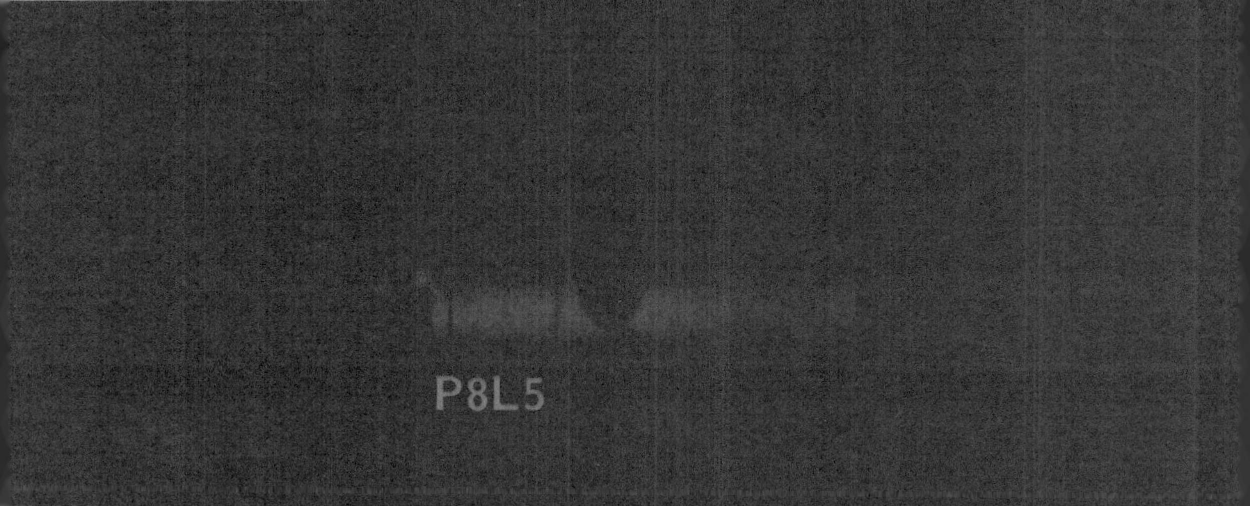

P8L5

A.a.3.5. Broken

25. P9L2

26. G3F2

27. G4F2

A.FUEL a.Pellets 3.Change of shape or location

P9L2

A.a.3.6. Dislocated and disintegrated

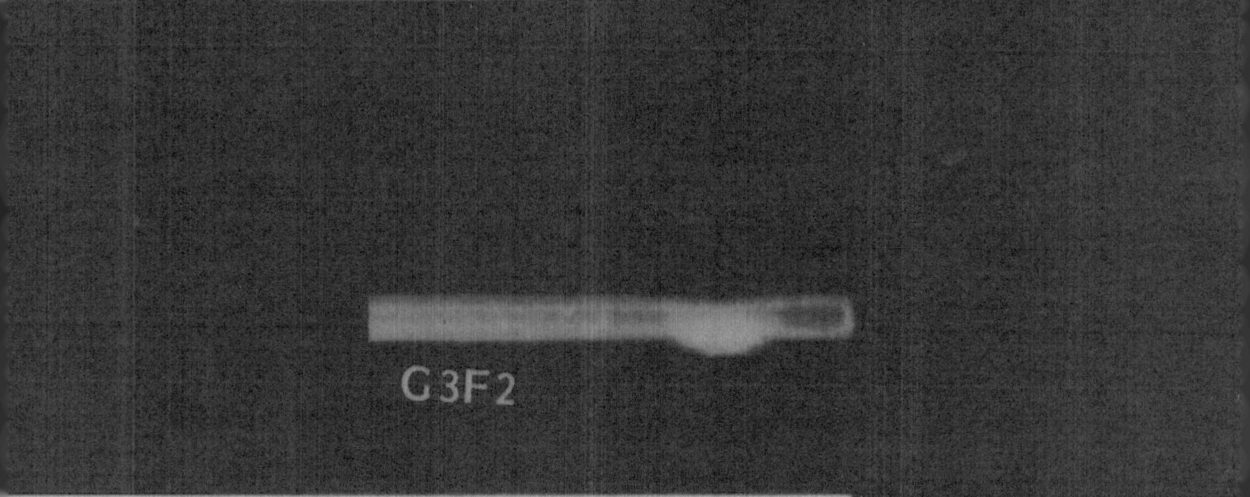

G3F2

A.a.3.8. Accumulated (fuel,outside of cladding)

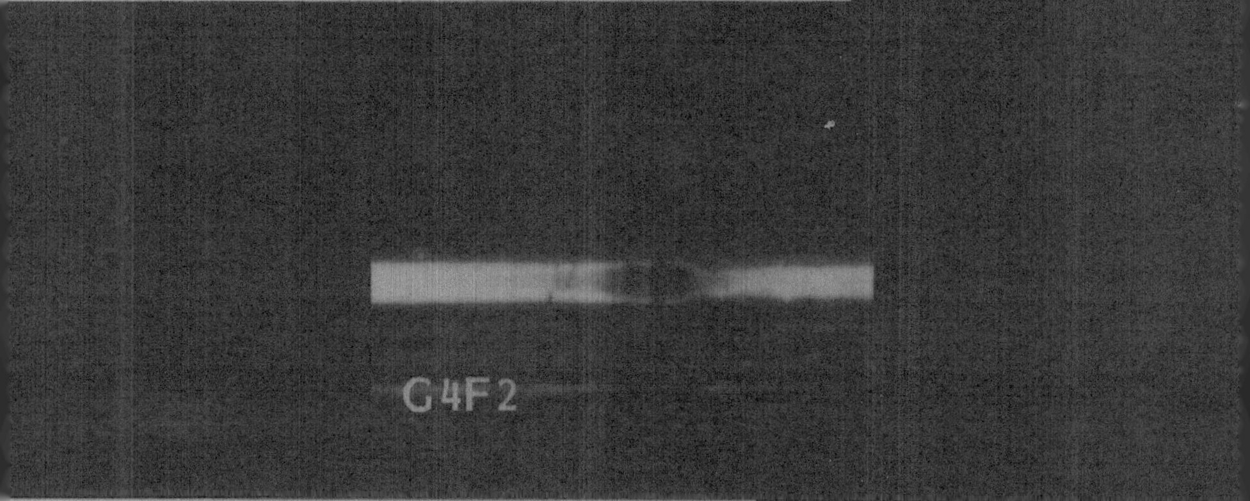

G4F2

A.a.3.9. Restructured (fuel,after melting)

28.

P10L5

29.

P11L5

30.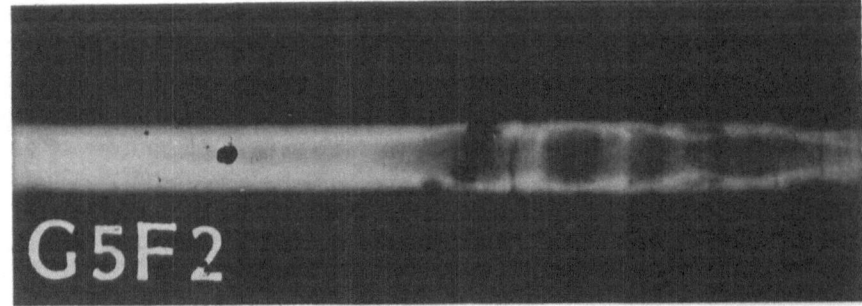

G5F2

A.FUEL a.Pellets 3.Change of shape or location

P10L5

A.a.3.9. Restructured (fuel, before restructuring)

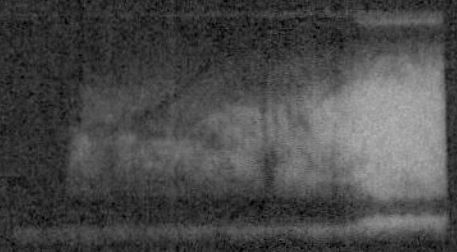

P11L5

A.a.3.9. Restructured (fuel, after restructuring)

G5F2

A.a.3.10.Melted (several pellets)

31.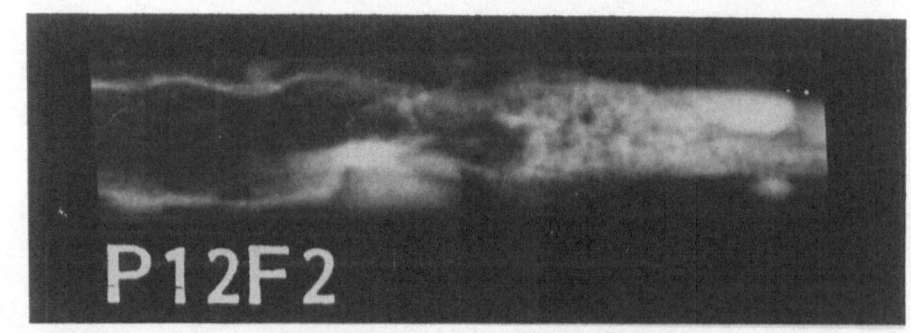

P12F2

32.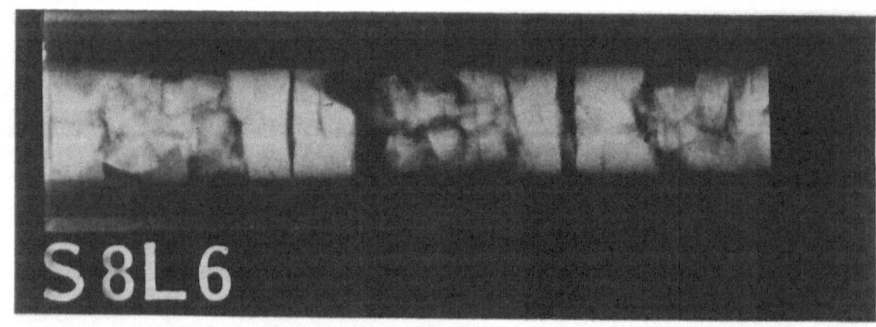

G6F2

33. S8L6

A.FUEL a.Pellets 3.Change of shape or location

P12F2

A.a.3.10.Melted

G6F2

A.a.3.11.Disintegrated

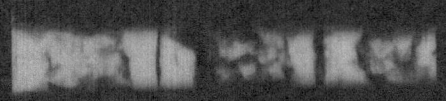

S8L6

A.a.3.11.Disintegrated

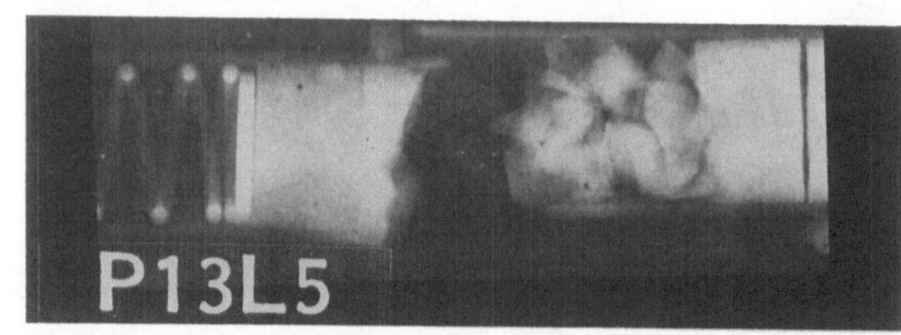

34.

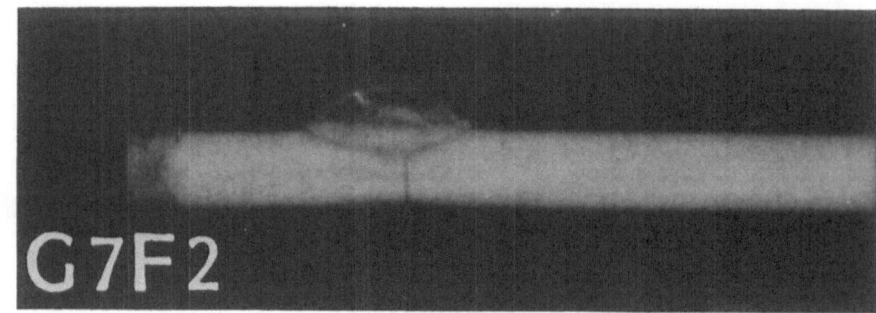

35.

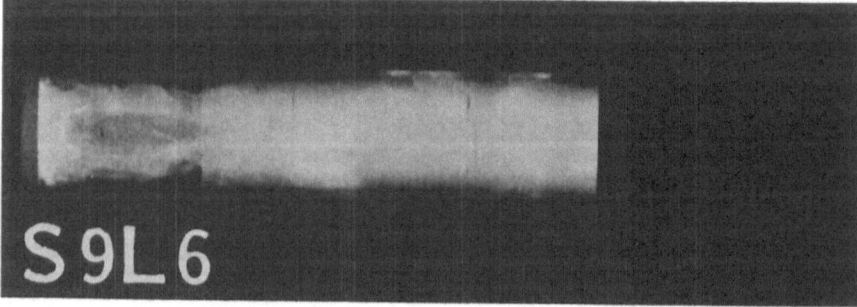

36.

A.FUEL a.Pellets 3.Change of shape or location

P13L5

A.a.3.11.Disintegrated

G7F2

A.a.3.12.Migration of fuel (outside of cladding)

A.FUEL a.Pellets 4.Voidage

S9L6

A.a.4.1. Central void in one pellet

37. 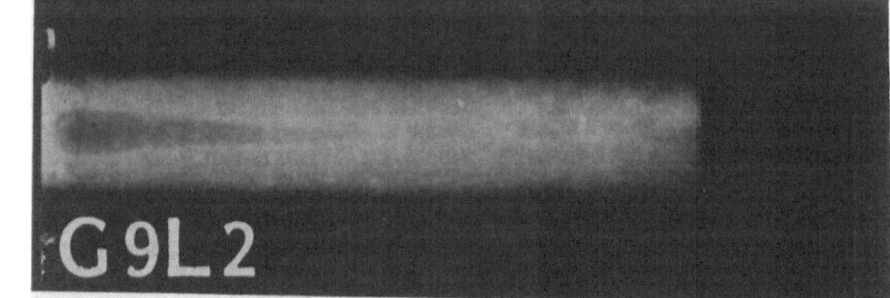 S10L6

38. G8L2

39. G9L2

A.FUEL a.Pellets 4.Voidage

S 10 L 6

A.a.4.1. Central void in one pellet

G 8 L 2

A.a.4.2. Central void (filled up) through several pellets

G 9 L 2

A.a.4.2. Central void (enlarged) through several pellets

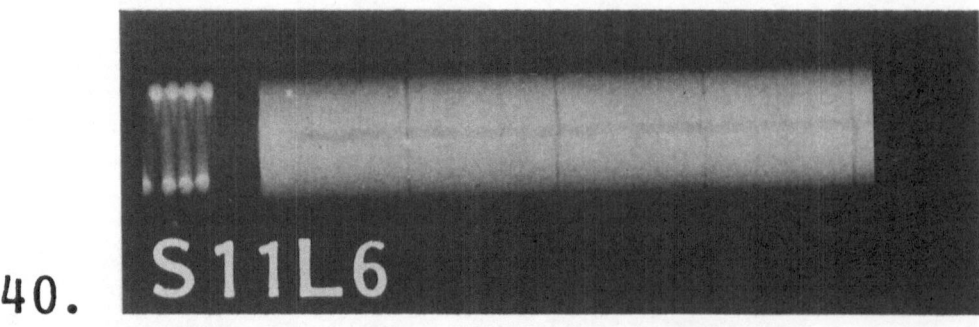

40.

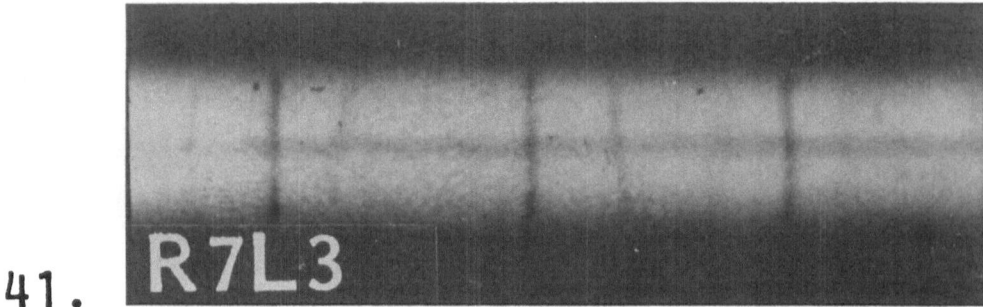

41.

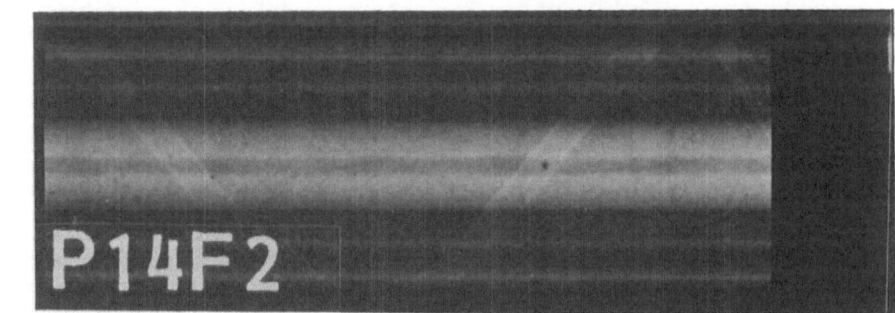

42.

A.FUEL a.Pellets 4.Voidage

S11L6

A.a.4.2. Central void through several pellets

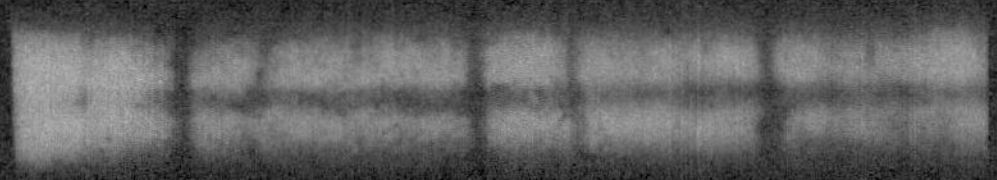

R7L3

A.a.4.3. Central void through fuel column

P14F2

A.a.4.3. Central void through fuel column

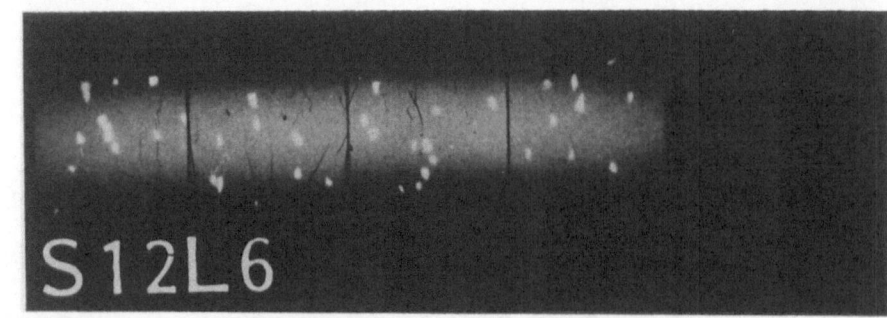

43. S12L6

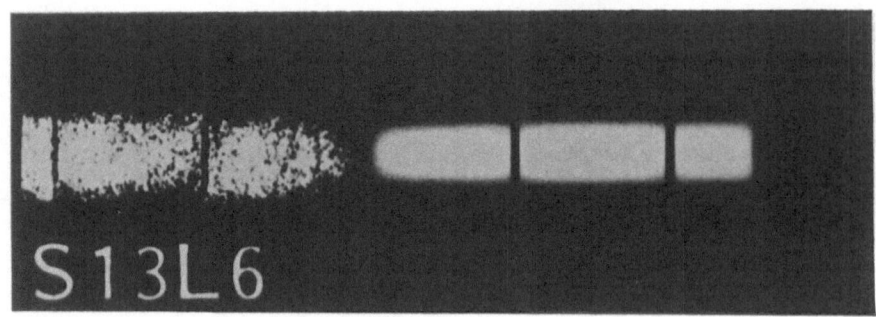

44. S13L6

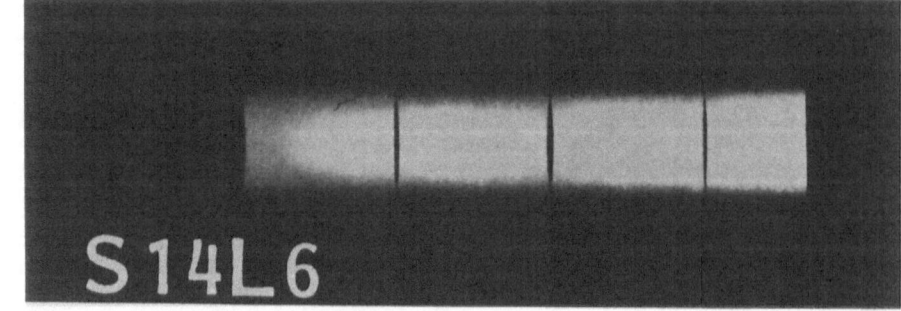

45. S14L6

A.FUEL a.Pellets 5.Inclusions

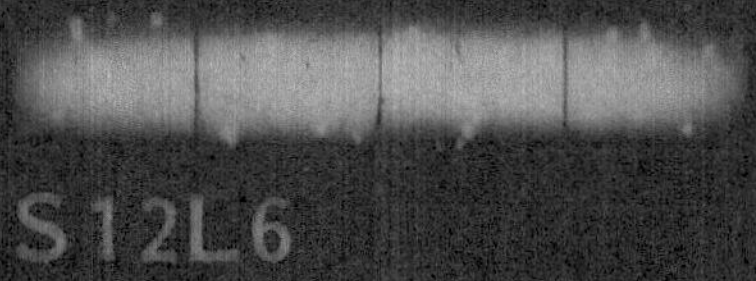

S12L6

A.a.5.1. Of Pu

S13L6

A.a.5.2. Of poison (left-granulated; right-powder)

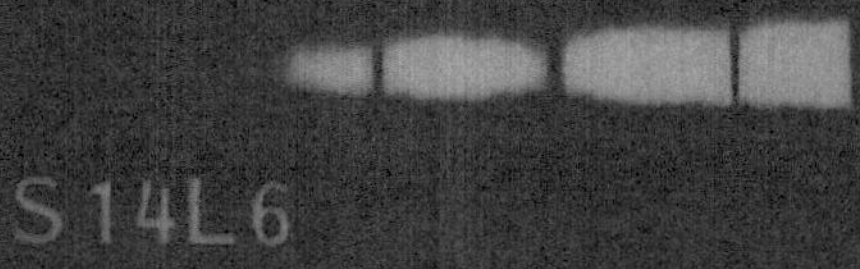

S14L6

A.a.5.2. Of poison (Gd_2O_2 powder, irradiated)

46.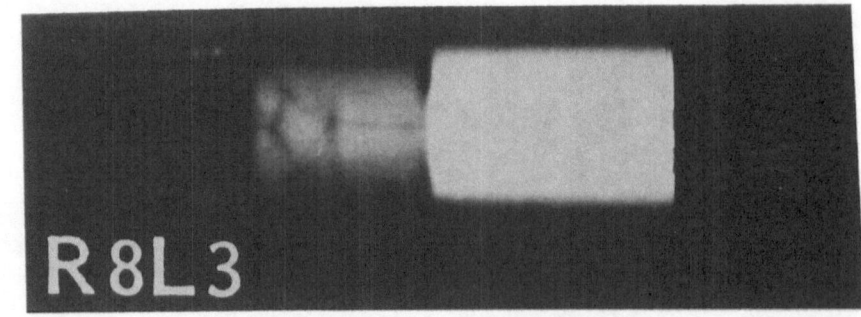

R8L3

47.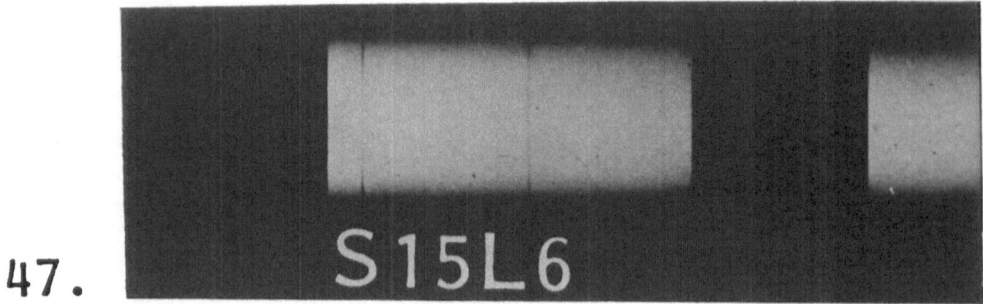

S15L6

48.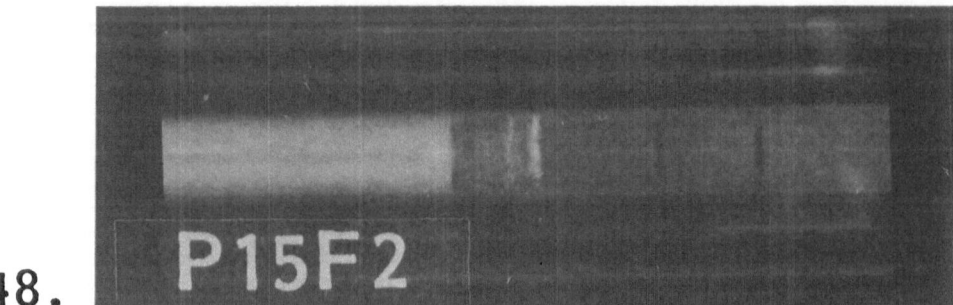

P15F2

R 8 L 3

S 15 L 6

P 15 F 2

49.

G 10 L 2

50.

P16L5

51.

P28L6

A.FUEL a.Pellets 8.Coolant

G10L2

A.a.8.1. In pellets

P16L5

A.a.8.1. In pellets (at dishing)

A.FUEL b.Annular 0.As fabricated

P28L6

A.b.0.

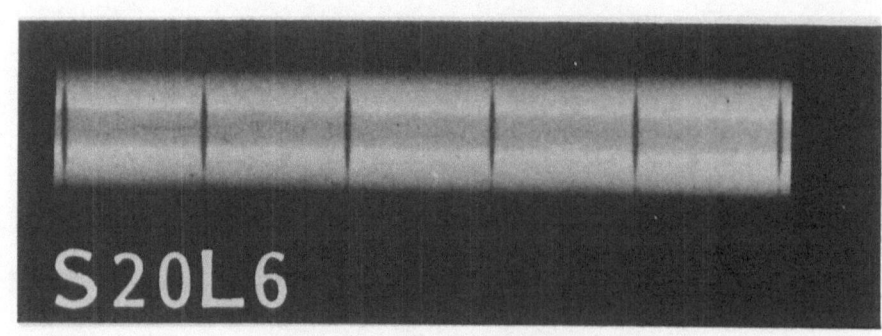

52.

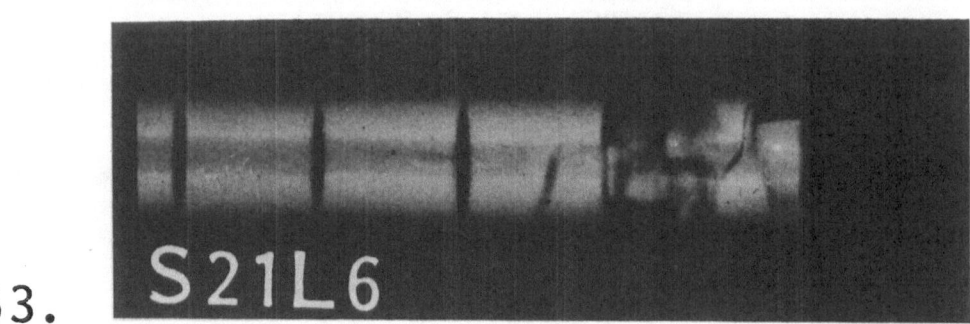

53.

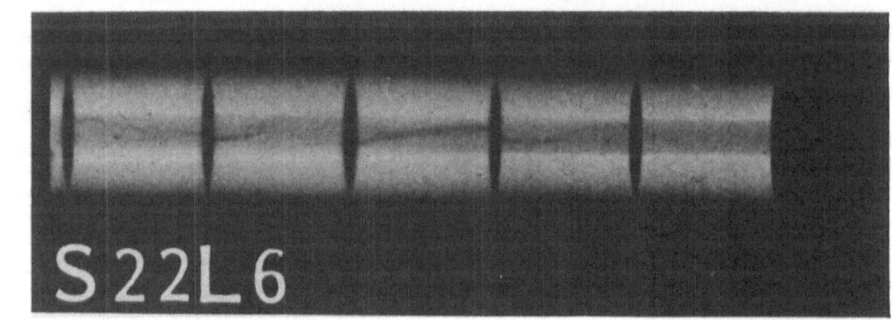

54.

A.FUEL b.Annular 0.As fabricated

S 20 L 6

A.b.0.

A.FUEL b.Annular 1.Cracks

S 21 L 6

A.b.1.1. Random

S 22 L 6

A.b.1.2. Londitudinal

55.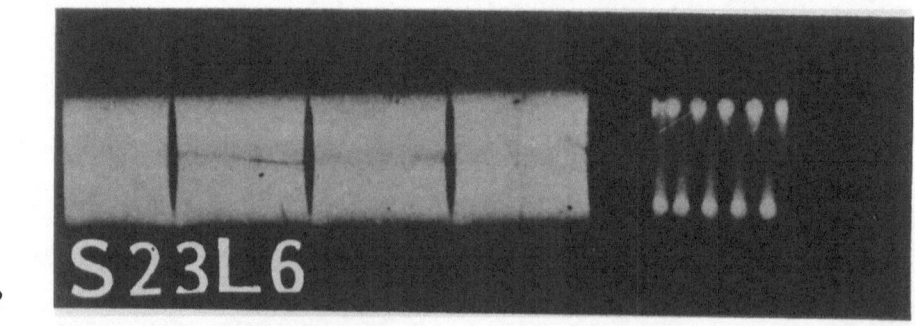

56.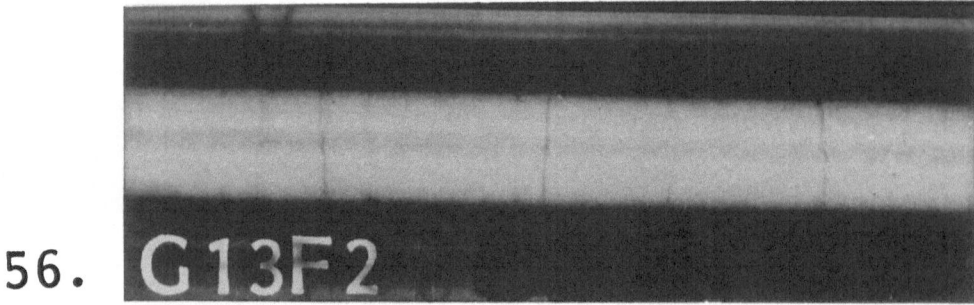

57.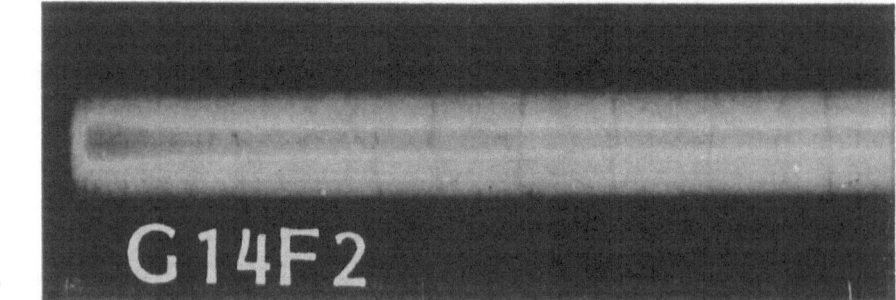

A.FUEL b.Annular 1.Cracks

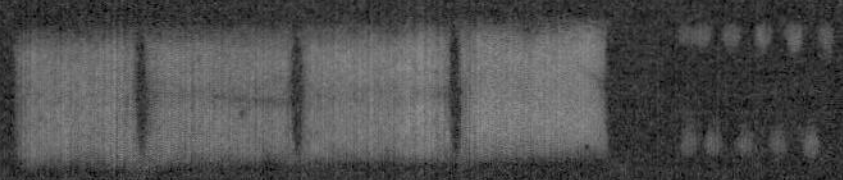

S23L6

A.b.1.3. Transverse

G13F2

A.b.1.3. Transverse

G14F2

A.b.1.5. Stratified

58.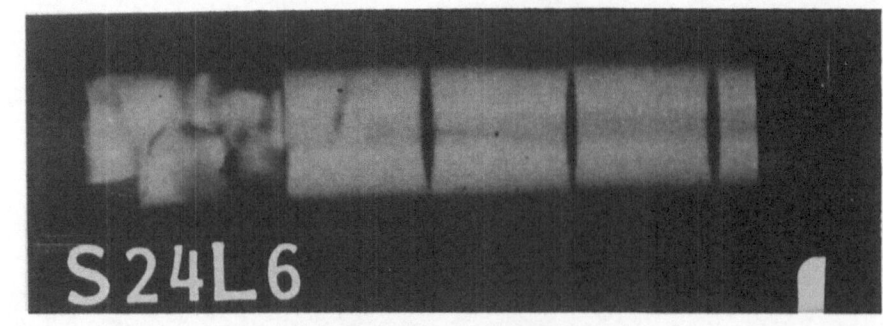

59.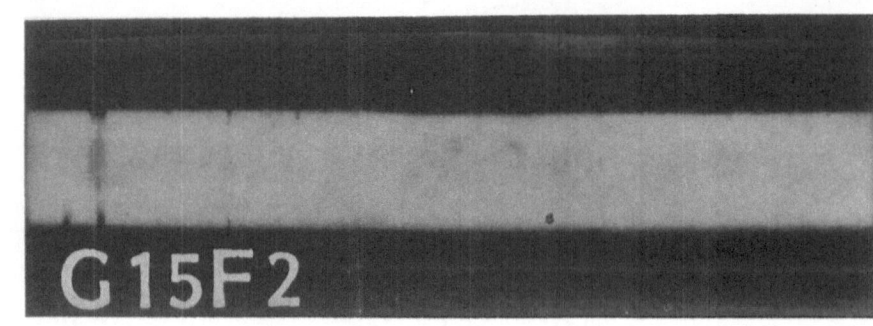

60.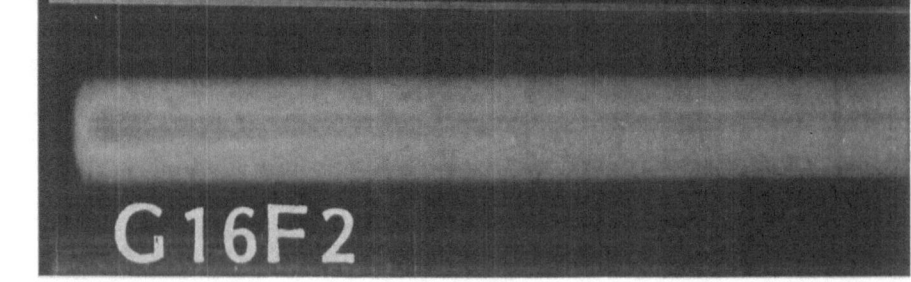

A.FUEL b.Annular 2.Chips

S24L6

A.b.2.2. Other

G15F2

A.b.2.4. Missing

A.FUEL b.Annular 3.Change of shape or location

G16F2

A.b.3.1. Central void enlarged

61.

62.

63.

A.FUEL b.Annular 3.Change of shape or location

G17F2

A.b.3.3. Central void filled-up (at several places)

G18F2

A.b.3.4. Central void deformed (excentric)

S25L6

A.b.3.4. Central void deformed

64.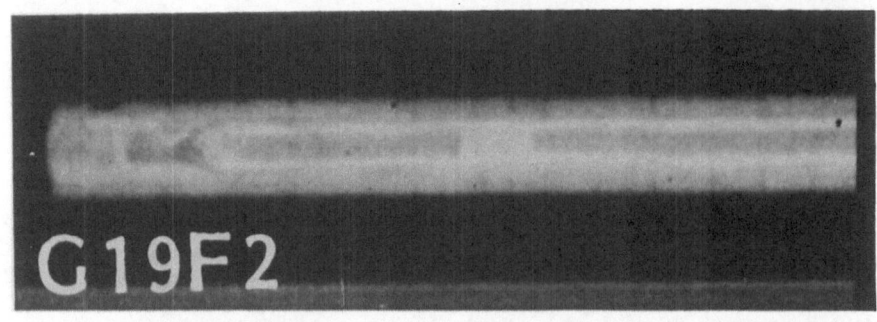

65.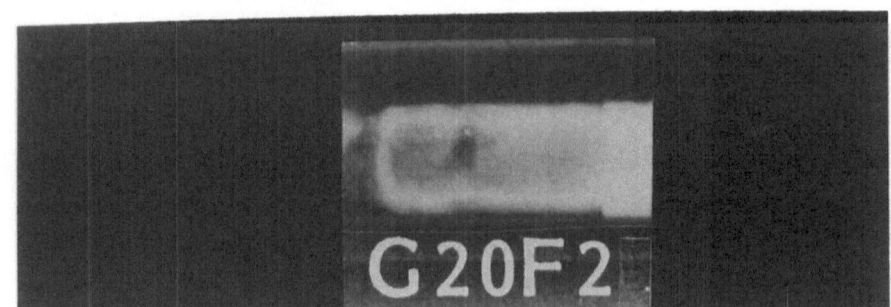

66.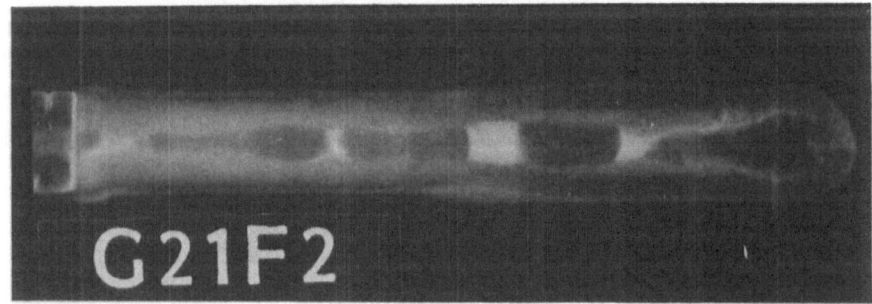

.FUEL b.Annular 3.Change of shape or location

G19F2

.b.3.8. Pu accumulated in central void

G20F2

.b.3.9. Restructuring (by initial defect)

G21F2

.b.3.10.Melting

67.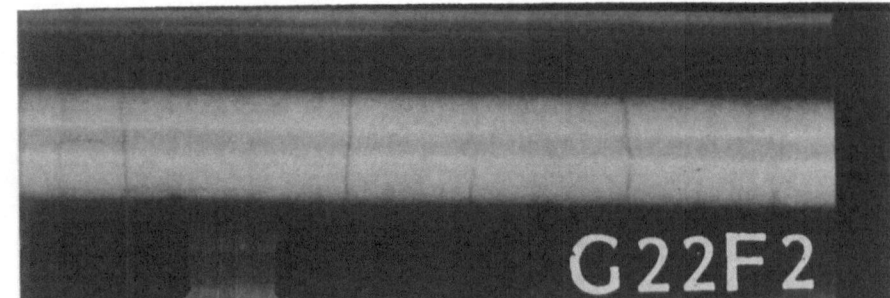

68.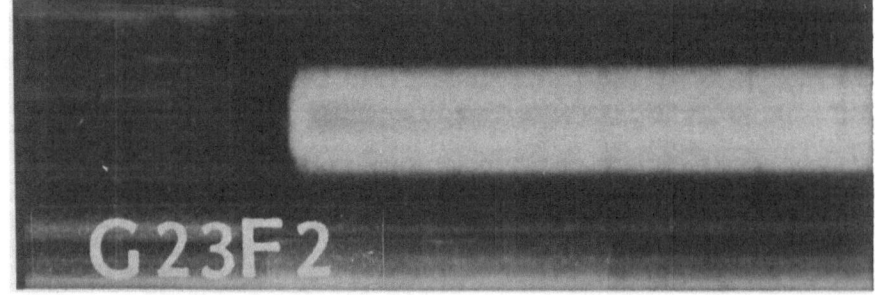

69.

A.FUEL b.Annular 3.Change of shape or location

S26L6

A.b.3.11.Disintegrated

G22F2

A.b.3.12.Migration of Pu (along central void)

A.FUEL b.Annular 4.Voidage

G23F2

A.b.4.3. Central void (increasing through several pellets)

70.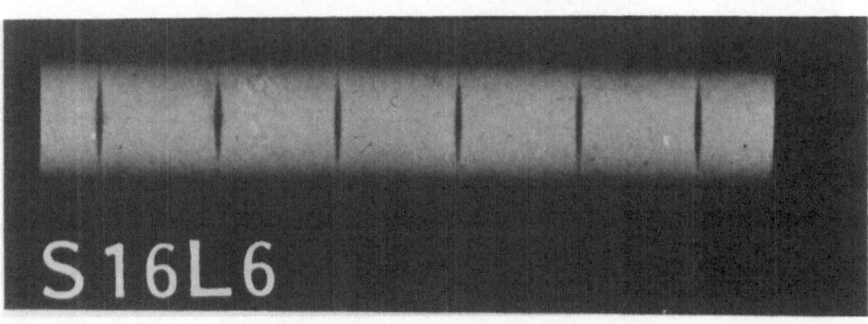

S16L6

71.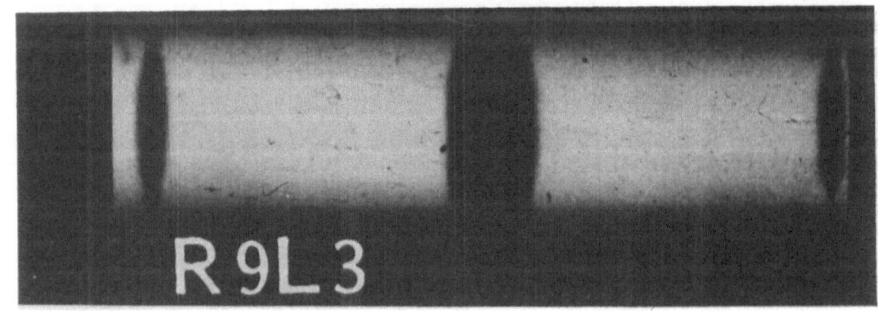

S17L6

72.

R9L3

A.FUEL c.Pellet-to-pellet gap 0.As fabricated

S16L6

A.c.0.

S17L6

A.c.0.

A.FUEL c.Pellet-to-pellet gap 3.Change of shape or location

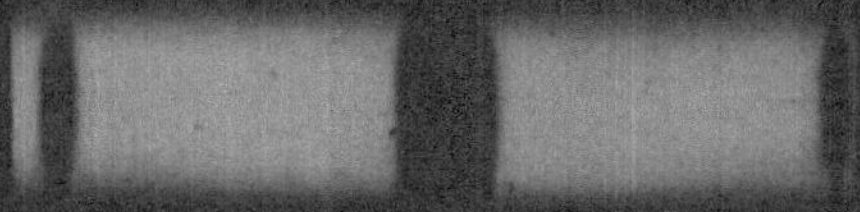

R9L3

A.c.3.1. Enlarged

73.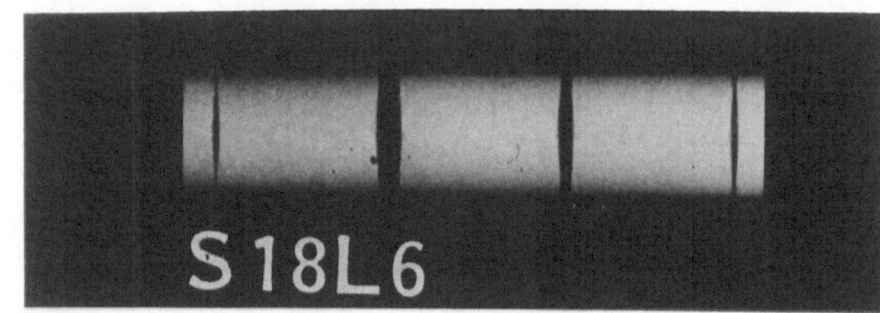

S 18 L 6

74.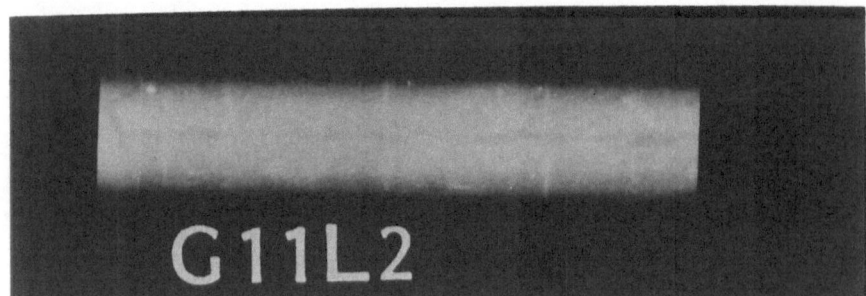

G 11 L 2

75.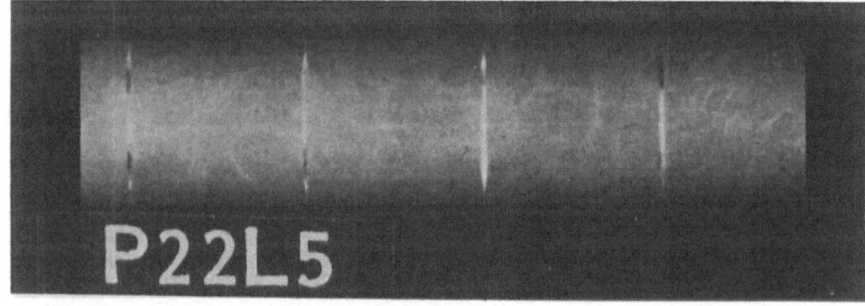

P 22 L 5

A.FUEL c.Pellet-to-pellet gap 3.Change of shape or location

S 18L6

A.c.3.1. Enlarged

A.FUEL c.Pellet-to-pellet gap 8.Coolant

G11L2

A.c.8.1. Filled with water (several gaps)

P22L5

A.c.8.1. Filled with water (original as P23L5)

76.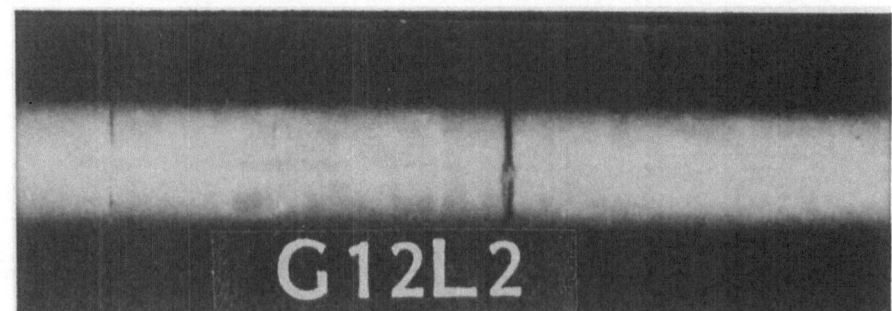

G12L2

77. P23L5

78.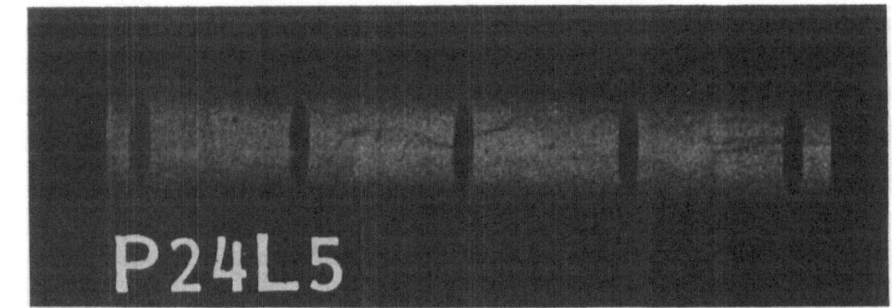

P24L5

A.FUEL c.Pellet-to-pellet gap 8.Coolant

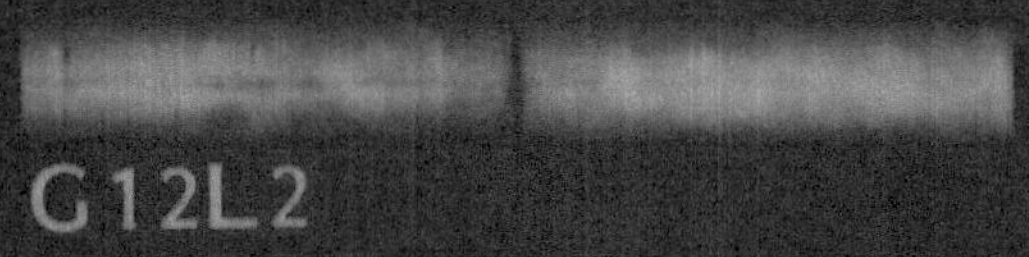

G12L2

A.c.8.2. Without water (one gap)

P23L5

A.c.8.2. Without water

A.FUEL d.Dishing 0.As fabricated

P24L5

A.d.0.

79. P25L2

80. P26L2

81. P27L5

A.FUEL d.Dishing 0.As fabricated

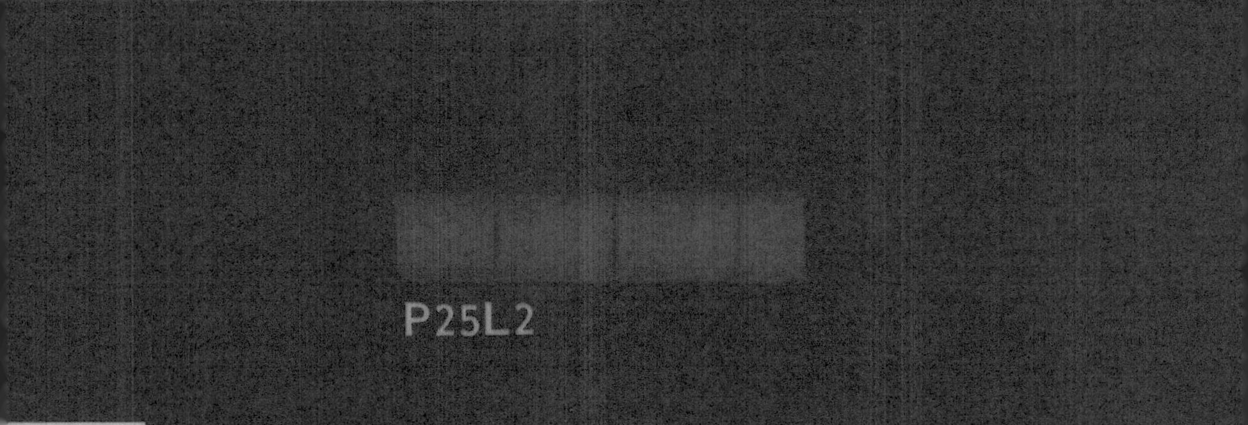

P25L2

A.d.0.

A.FUEL d.Dishing 3.Change of shape or location

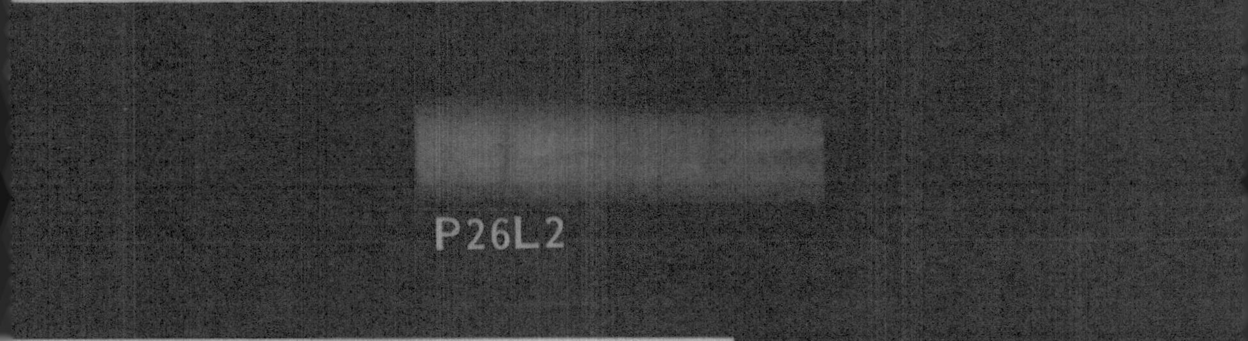

P26L2

A.d.3.3. Disappeared (original as P25L2)

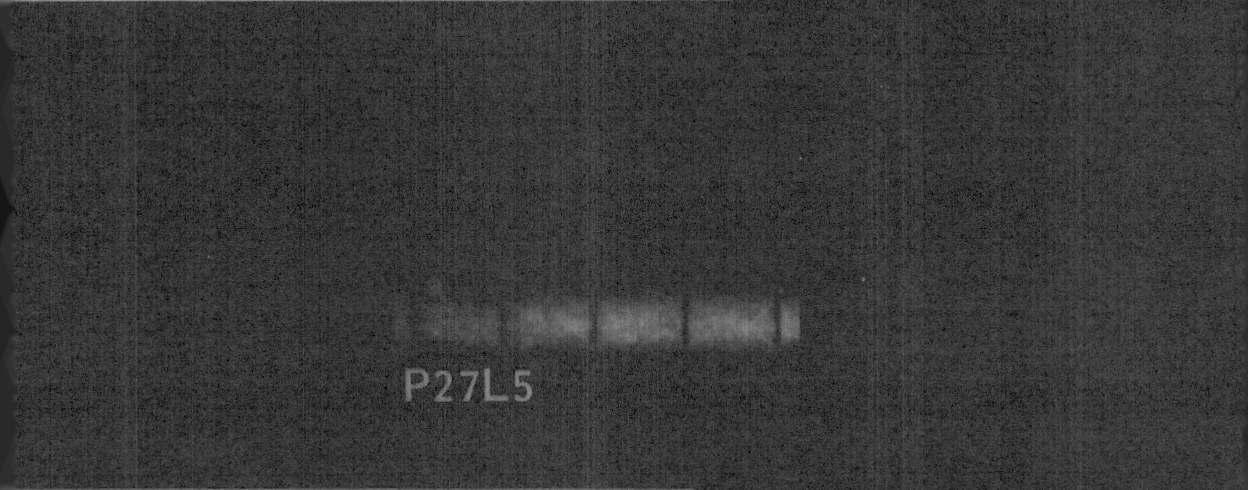

P27L5

A.d.3.4. Deformed (original as P24L5)

82.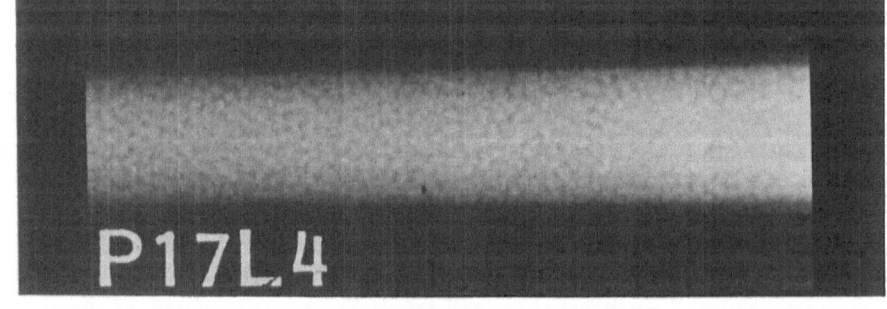

R10L3

83.

S19L6

84.

P17L4

A.FUEL d.Dishing 3.Change of shape or location

R10L3

A.d.3.4. Deformed

S19L6

A.d.3.4. Deformed

A.FUEL e.Vibro-compacted 0.As fabricated

P17L4

A.e.0. (sol-gel)

85. P18L2

86. P19L2

87. P20L4

A.FUEL e.Vibro-compacted 0.As fabricated

P18L2

A.e.0. (kernels)

A.FUEL e.Vibro-compacted 1.Cracks

P19L2

A.e.1.3. Transverse

A.FUEL e.Vibro-compacted 2.Chips

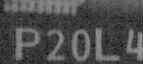

P20L4

A.e.2.4. Missing

88.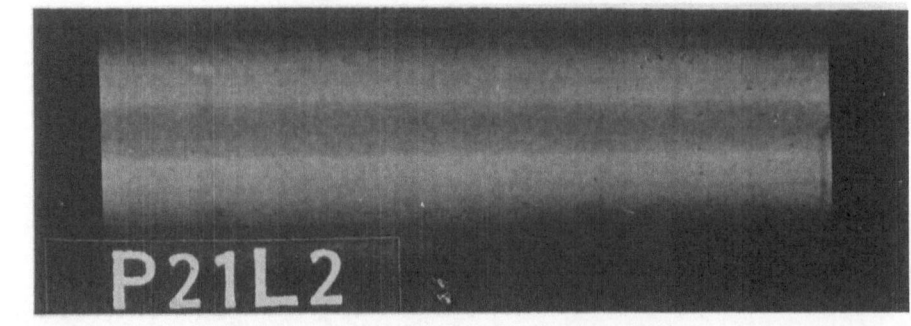

P21L2

89. S27F6

90.

S28L6

A.FUEL e.Vibro-compacted 4.Voidage

P21L2

A.e.4.3. Central void through column

A.FUEL f.Fuel-to-clad gap 0.As fabricated

S 27F6

A.f.0.

S 28L6

A.f.0.

91.

92.

93.

A.FUEL f.Fuel-to-clad gap 0.As fabricated

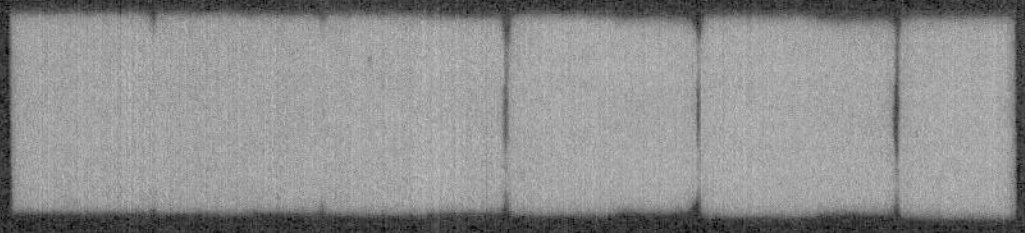

R21L4

A.f.0. (calibrated gap)

A.FUEL f.Fuel-to-clad gap 3.Change of shape or location

R11L3

A.f.3.1. Enlarged

G24L2

A.f.3.1. Enlarged

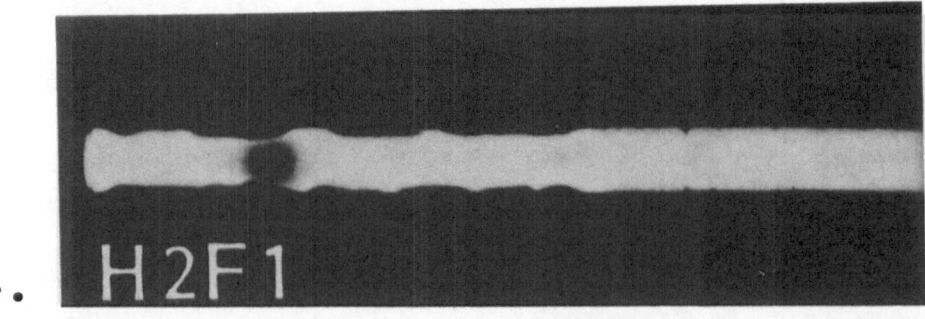

94. H2F1

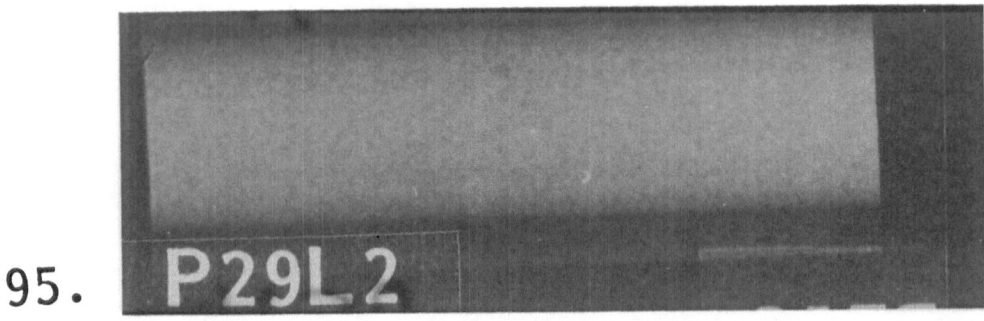

95. P29L2

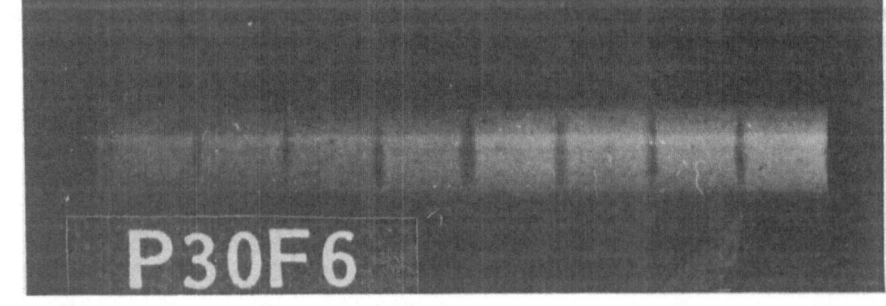

96. P30F6

A.FUEL f.Fuel-to-clad gap 3.Change of shape or location

H 2F 1

A.f.3.4. Deformed

A.FUEL g.Fuel column 0.As fabricated

P29L 2

A.g.0. (vibro-compacted)

P30F6

A.g.0. (pelletized)

97.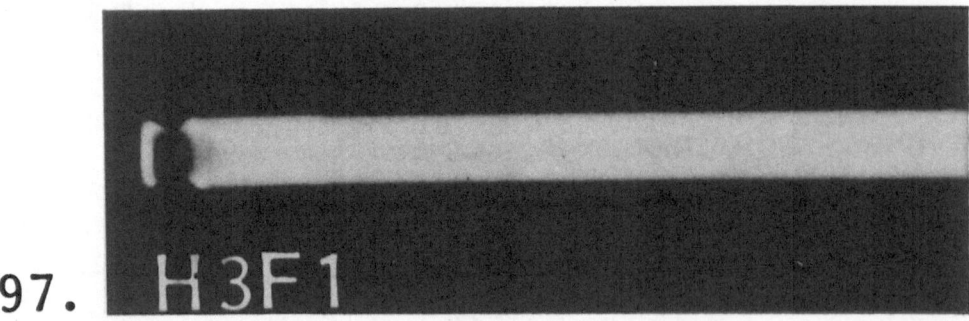

98.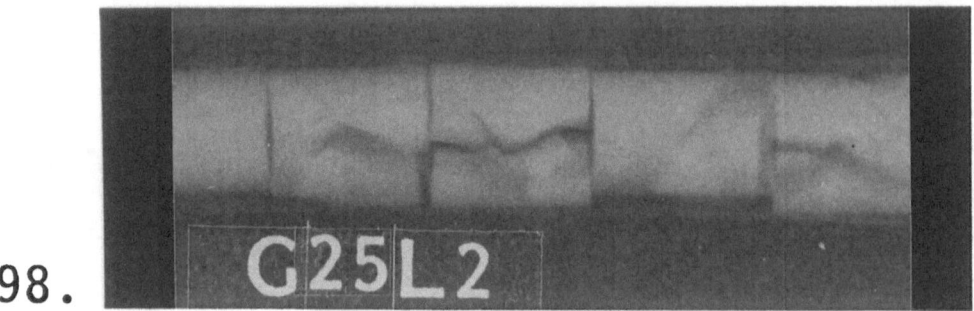

99.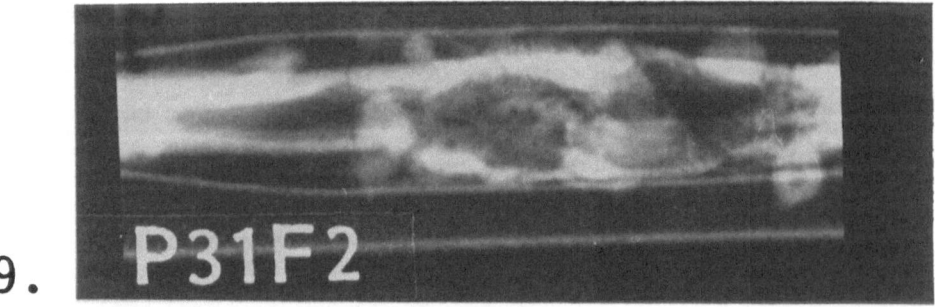

A.FUEL g.Fuel column 3.Change of chape or location

H 3F 1

A.g.3.2. Contracted

G25L 2

A.g.3.4. Deformed

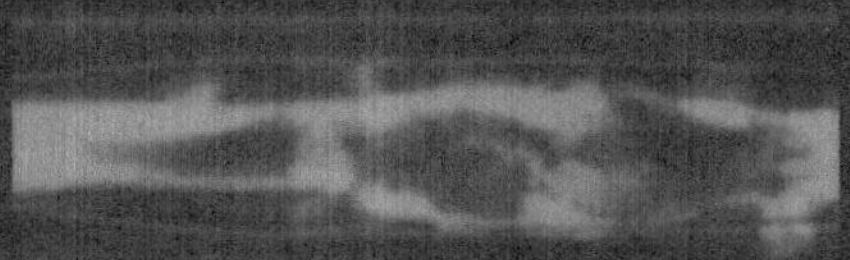

P31F2

A.g.3.4. Deformed (fuel melted)

100. P32F2

101. G26F2

102. G27F2

A.FUEL g.Fuel column 3.Change of shape or location

P32F2

A.g.3.4. Deformed (stack repartition)

G26F2

A.g.3.7. Extended (by melting)

G27F2

A.g.3.10.Melted (whole column)

103.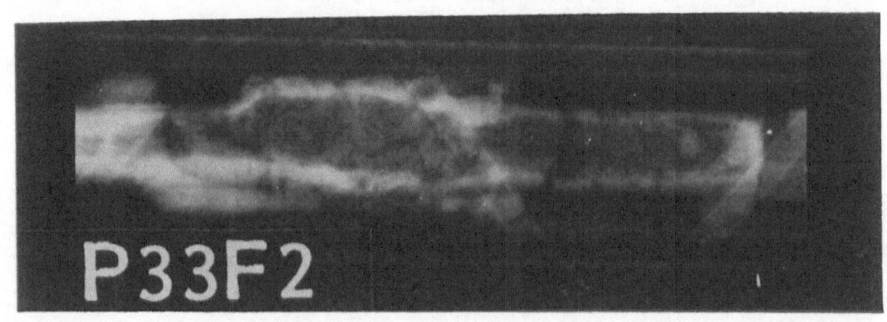

P33F2

104.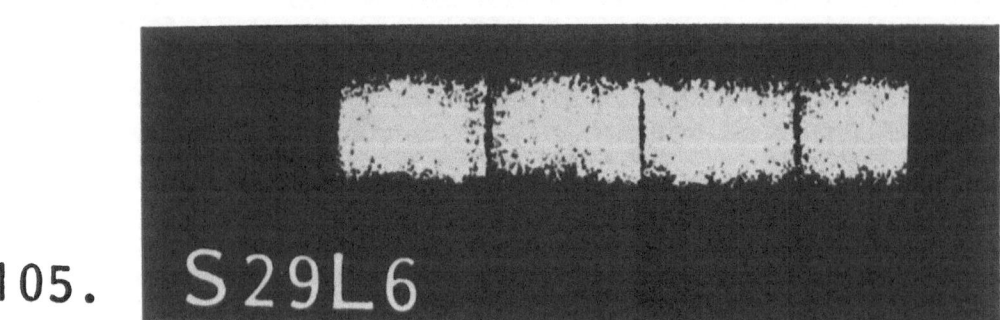

P34F2

105.

S 29 L 6

A.FUEL g.Fuel column 3.Change of shape or location

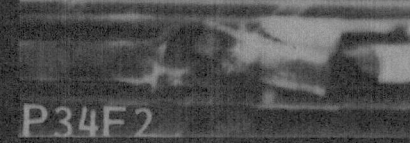

P33F2

A.g.3.10.Melted (and partly disappeared)

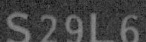

P34F2

A.g.3.11.Disintegrated

A.FUEL h.Composition 0.As fabricated

S29L6

A.h.0.   (with granular Gd poison)

106.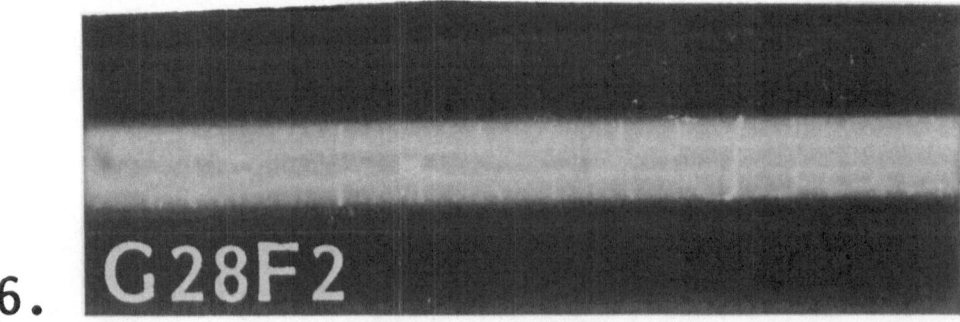

G28F2

107.

G29F2

108.

G30F2

A.FUEL h.Composition 3.Change of shape or location

G28F2

A.h.3.12.Migration of Pu (in transverse cracks)

G29F2

A.h.3.12.Migration of Pu (along central void)

G30F2

A.h.3.12.Migration of Pu (along central void)

109.

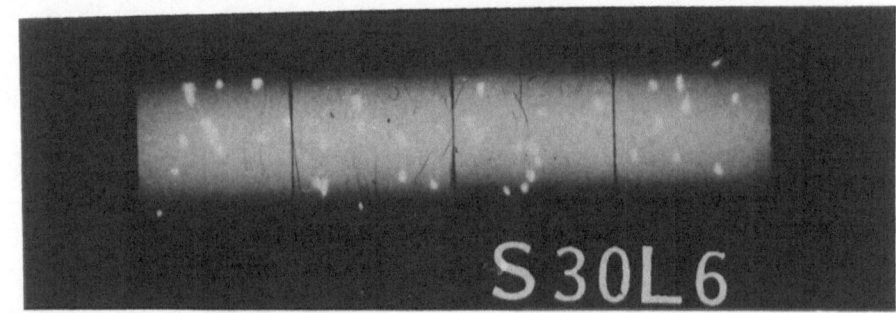

S 30 L 6

110.

S 31 L 6

111.

P 35 L 4

A.FUEL h.Composition 5.Inclusions

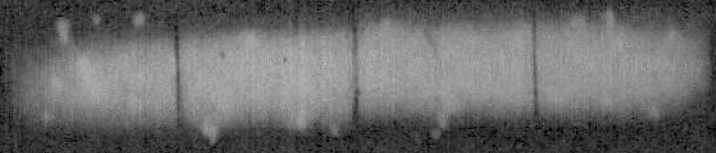

S 30L 6

A.h.5.1. Of Pu

S 31L 6

A.h.5.2. Of poison

B.CLADDING 0.As fabricated

P 35 L 4

B.0.   (PWR fuel rod)

112.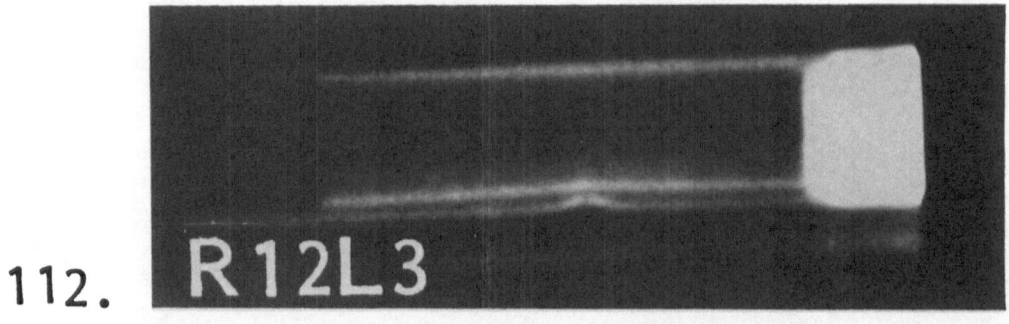

R12L3

113.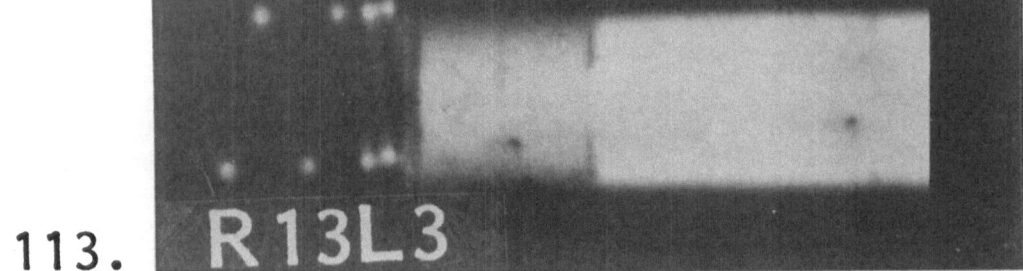

R13L3

114.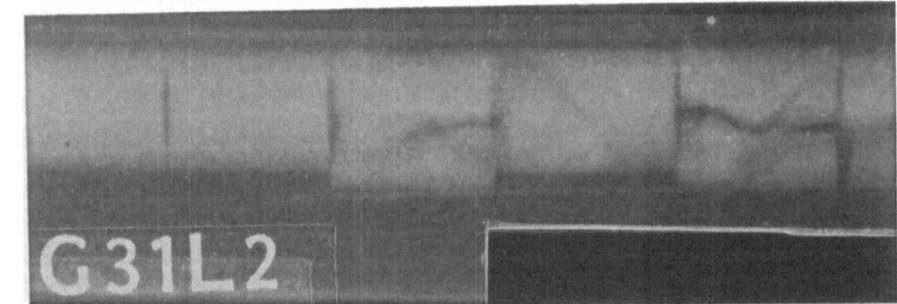

G31L2

B.CLADDING 3.Change of shape or location

R12L3

B.3.4. Deformed

R13L3

B.3.5. Broken

G31L2

B.3.5. Broken

115.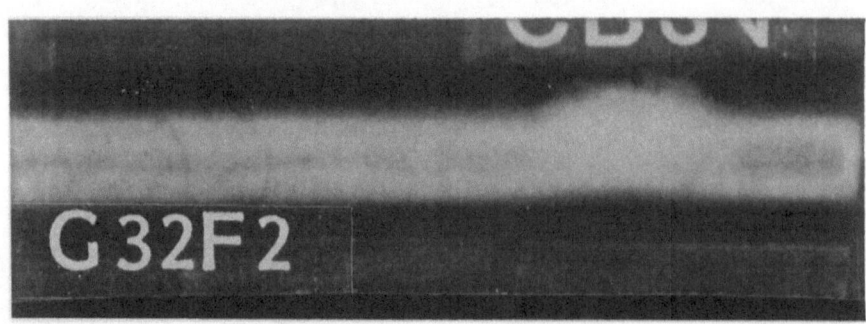

116.

117.

B.CLADDING 3.Change of shape or location

G 32F 2

B.3.5. Broken

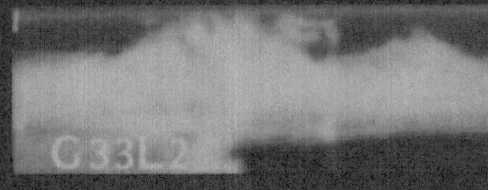

G 33L 2

B.3.5. Broken

G 34L 2

B.3.7. Extended

118.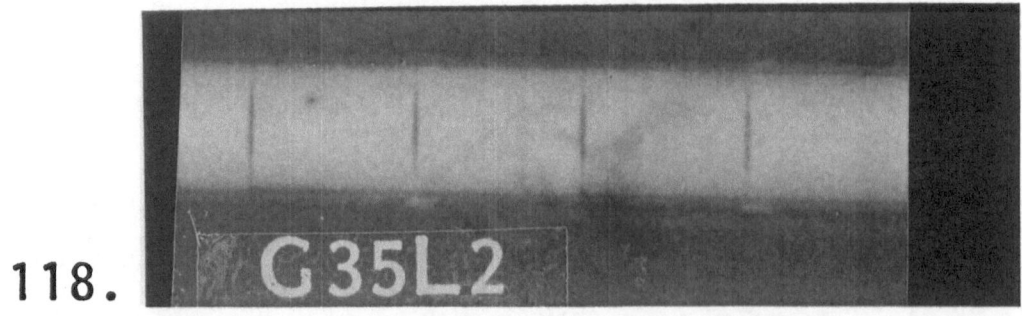

119.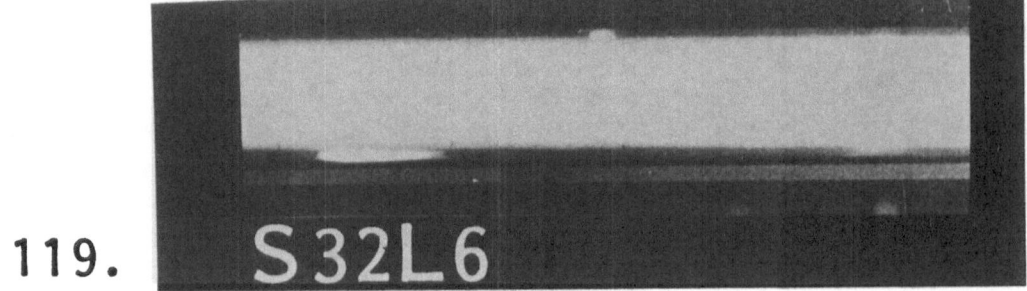

120.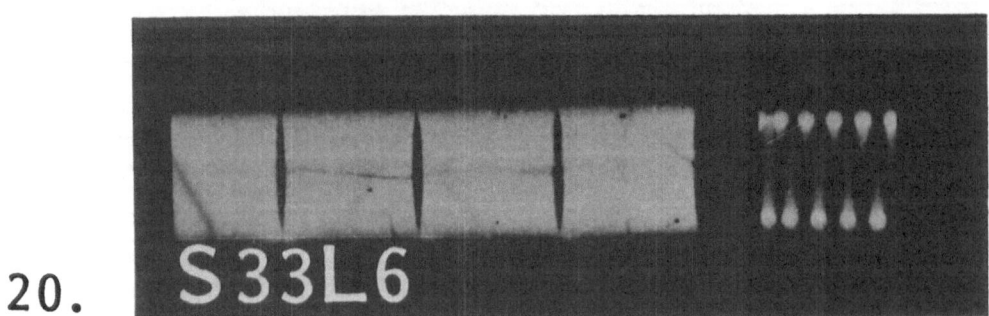

G35L2

B.6.1. Hydrides (in front of pellet-to-pellet gaps)

S32L6

B.6.1. Hydrides

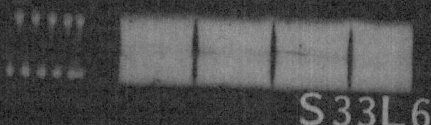

S33L6

B.6.1. Hydrides

121.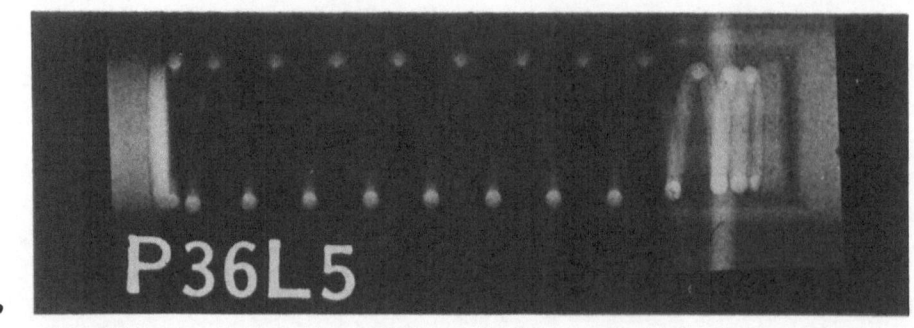

P36L5

122.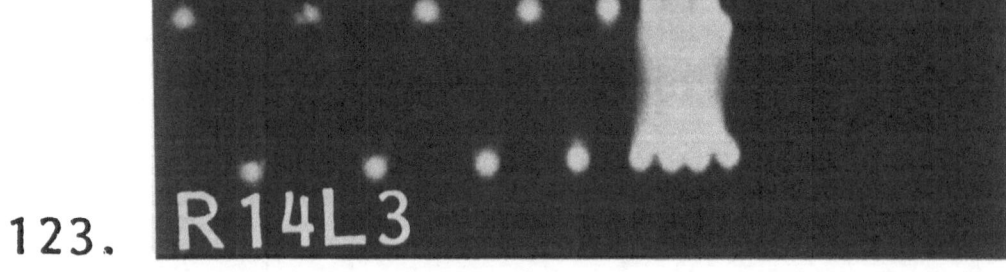

P37L2

123. R14L3

C.PLENUM a.Spring 0.As fabricated

P36L5

C.a.0. (BWR test fuel rod)

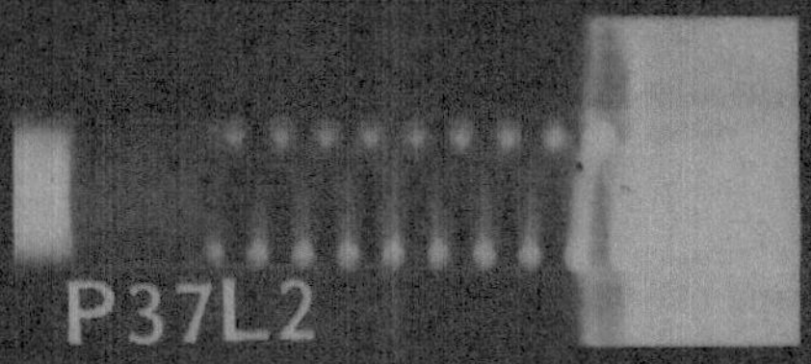

P37L2

C.a.0. (PWR test fuel rod)

C.PLENUM a.Spring 3.Change of shape or location)

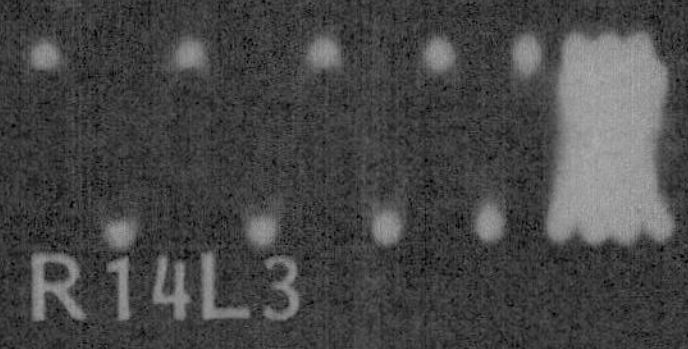

R14L3

C.a.3.2. Contracted

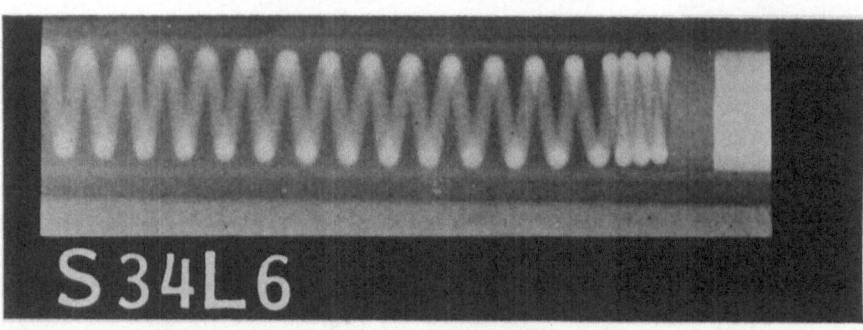

124.

S 34 L 6

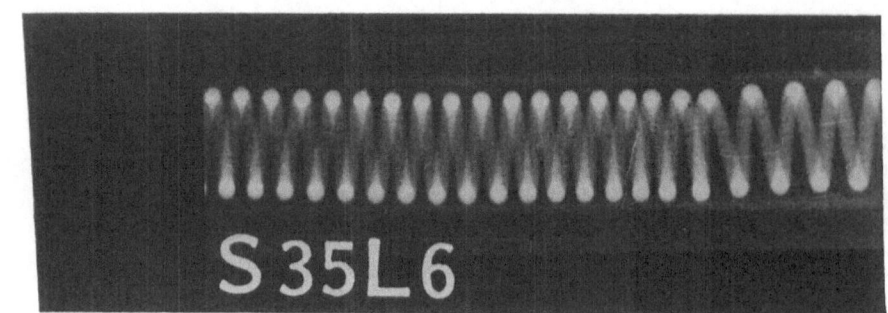

125.

S 35 L 6

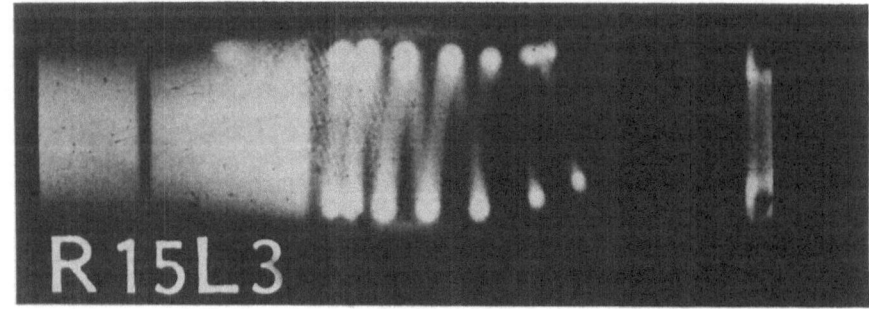

126.

R 15 L 3

C.PLENUM a.Spring 3.Change of shape or location

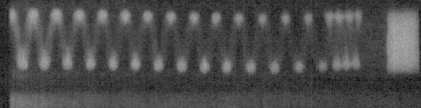

S34L6

C.a.3.2. Contracted

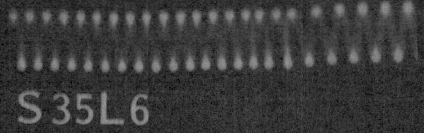

S35L6

C.a.3.4. Deformed

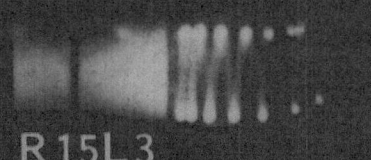

R15L3

C.a.3.6. Dislocated

127.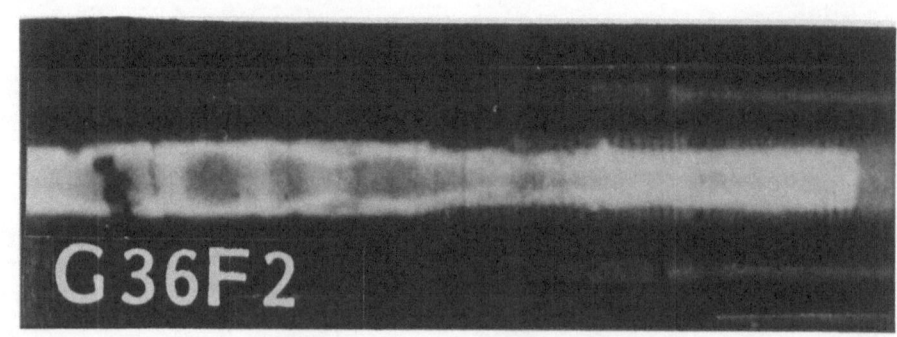

128.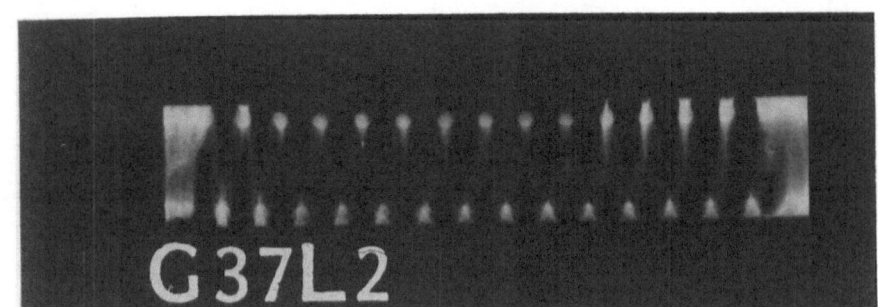

129.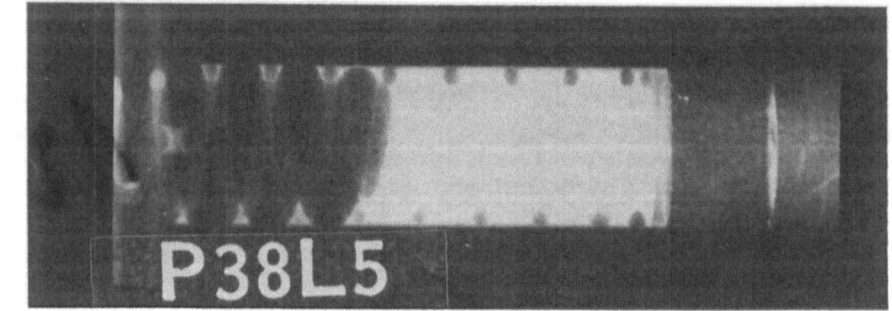

C.PLENUM a.Spring 3.Change of shape or location

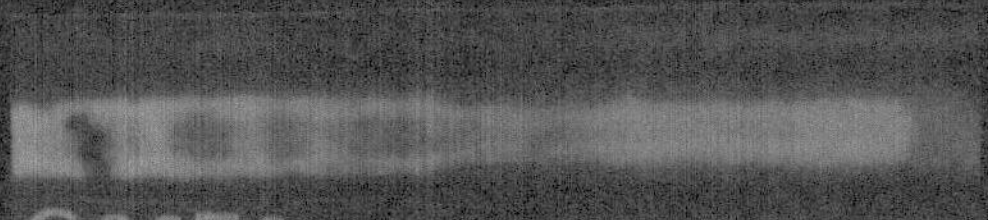

G 36F 2

C.a.3.11.Disintegrated (by melting)

C.PLENUM a.Spring 8.Coolant

G 37L 2

C.a.8.1. On Spring

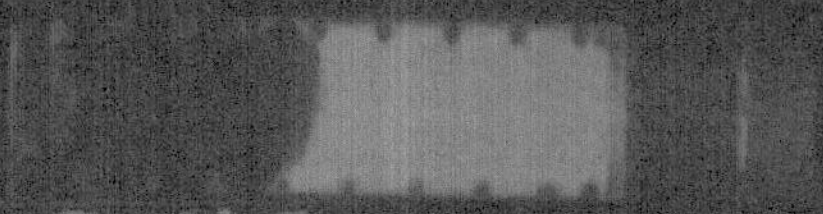

P 38L 5

C.a.8.1. In spring (flooded)

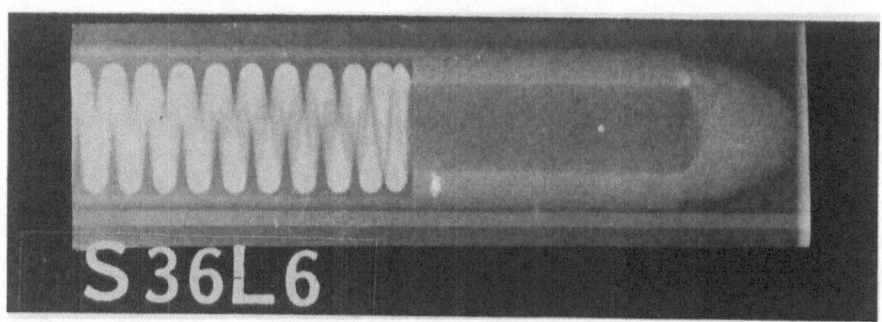

130. S36L6

131. R16L3

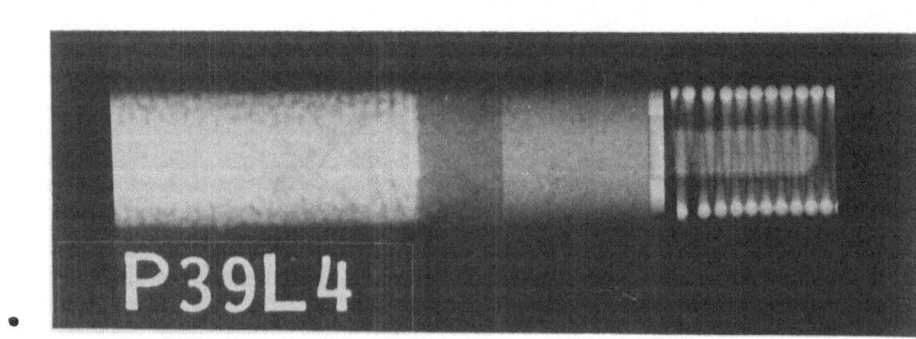

132. P39L4

C.PLENUM b.Spring sleeve 0.As fabricated

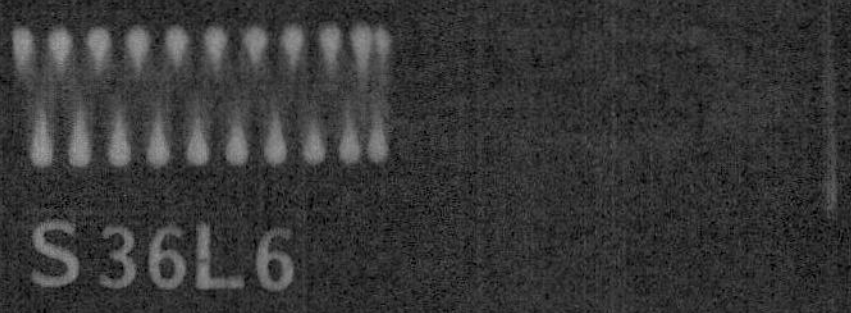

S 36L 6

C.b.0.

C.PLENUM b.Spring sleeve 3.Change of shape or location

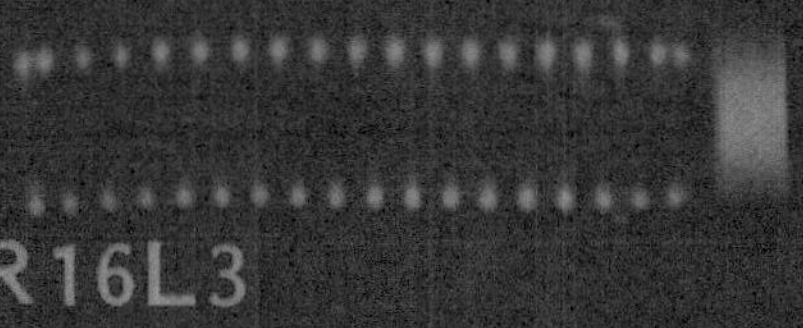

R 16L 3

C.b.3.5. Broken

C.PLENUM c.Insulating disc 0.As fabricated

P 39L 4

C.c.0.   (top of vibro-compacted fuel)

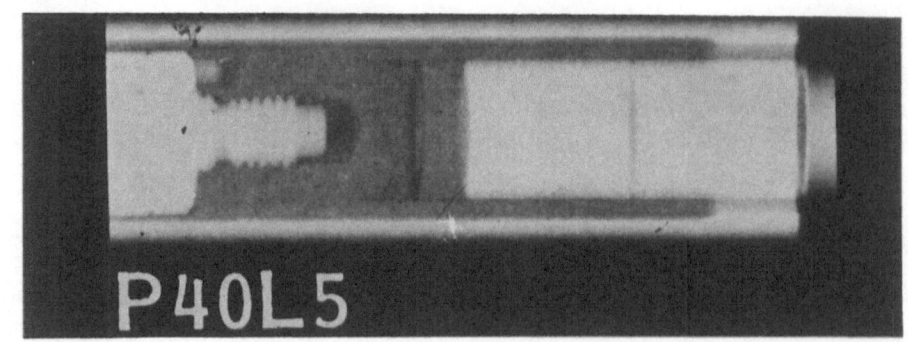

133. P40L5

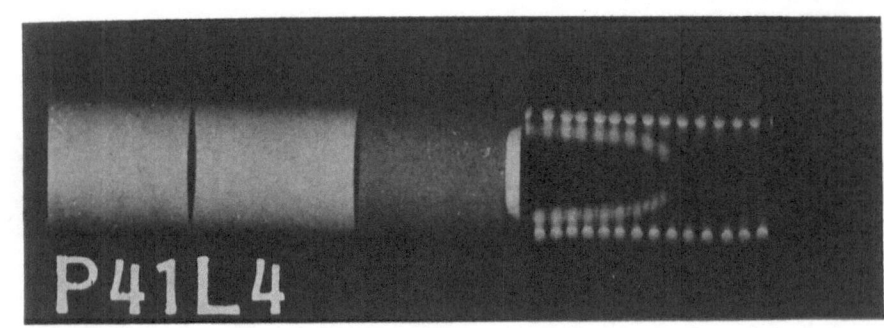

134. P41L4

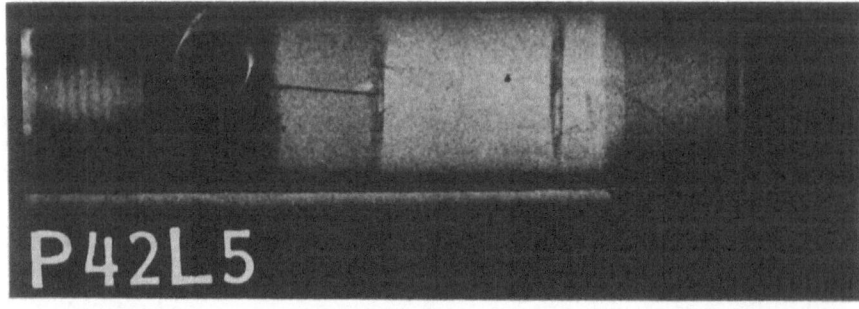

135. P42L5

C.PLENUM c.Insulating disc 0.As fabricated

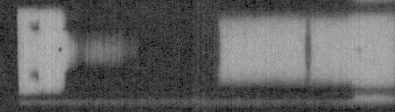

P40L5

C.c.0.   (bottom of pelletized fuel)

P41L4

C.c.0.   (top of pelletized fuel)

C.PLENUM c.Insulating disc 3.Change of shape or location

P42L5

C.c.3.5. Broken

136.

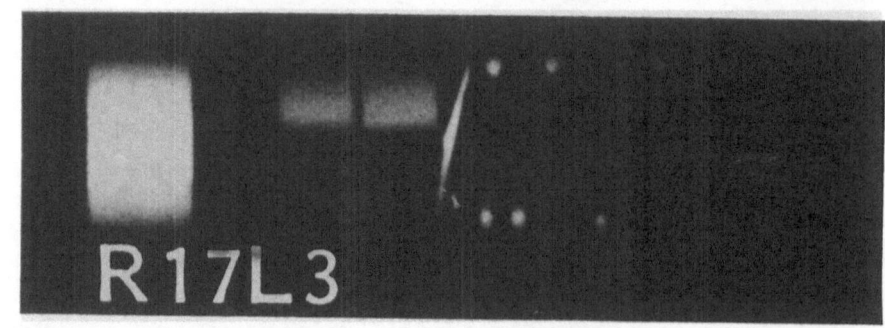

R 17L 3

137.

G 38L 2

138.

S 37F 6

C.PLENUM c.Insulating disc 3.Change of shape or location

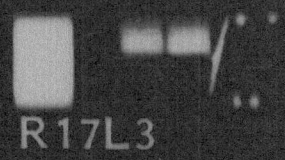

R17L3

C.c.3.6. Dislocated

C.PLENUM c.Insulating disc 6.Corrosion

G38L2

C.c.6.1. Hydrides (in Al_2O_3)

C.PLENUM d.Spacer 0.As fabricated

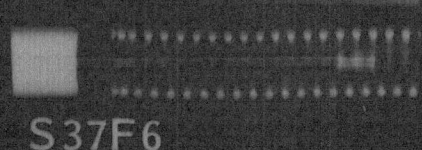

S37F6

C.d.0.

139.

S38L6

140.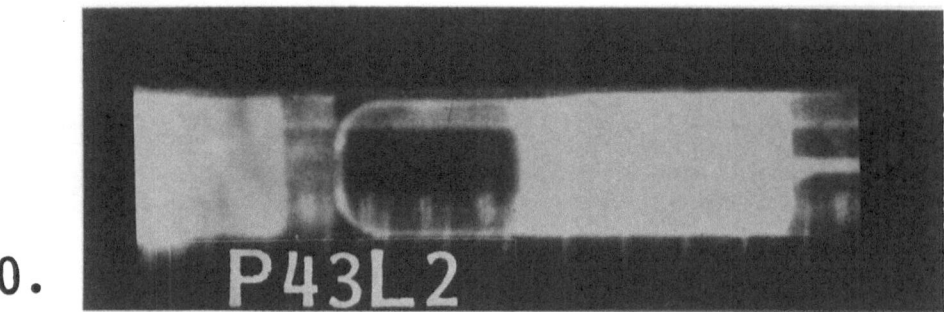

P43L2

141.

P44L2

C.PLENUM d.Spacer 0.As fabricated

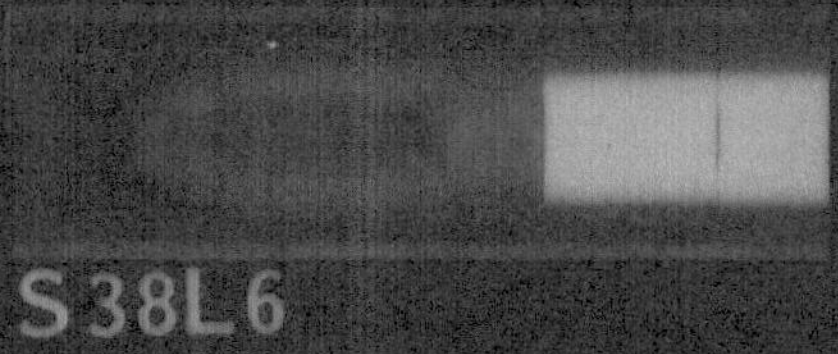

S38L6

C.d.0.

C.PLENUM d.Spacer 8.Coolant

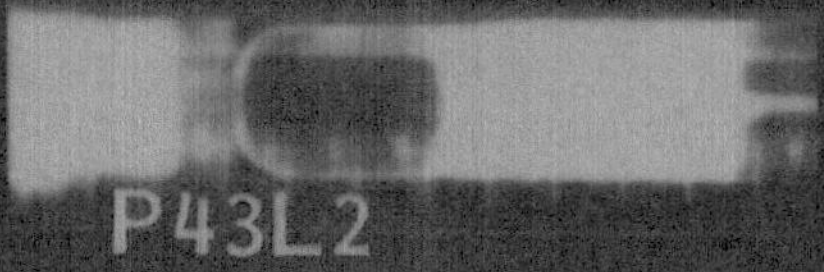

P43L2

C.d.8.1. In spacer (at bottom)

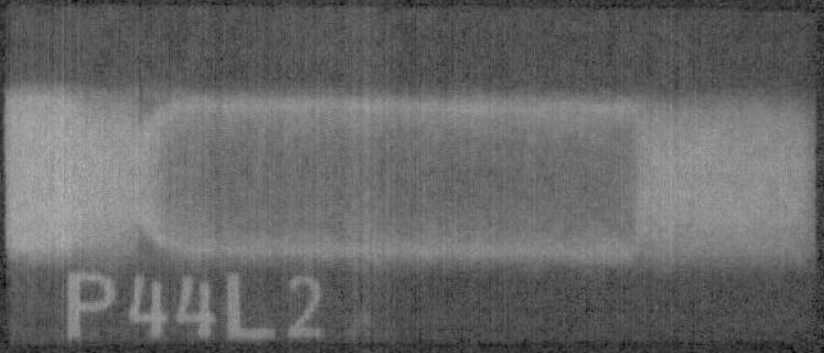

P44L2

C.d.8.2. Absent (P42L2 not irradiated)

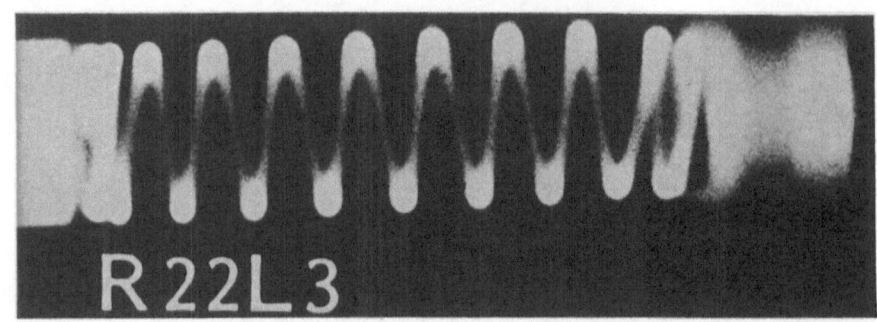

142. R 22 L 3

143. G 39 L 2

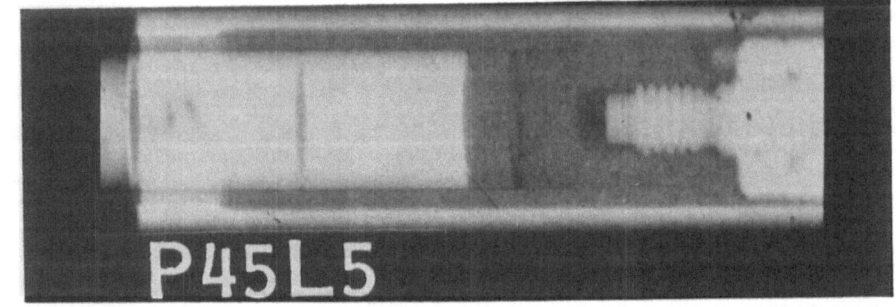

144. P 45 L 5

C.PLENUM e.Fuel column-to-plug 0.As fabricated

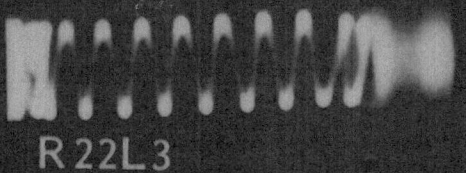

R 22L3

C.e.0.

C.PLENUM e.Fuel column-to-plug 8.Coolant

G39L2

C.e.8.1. In plenum

D.PLUG a.Bottom 0.As fabricated

P45L5

D.a.0.   (PWR test fuel rod)

145.

S39L6

146.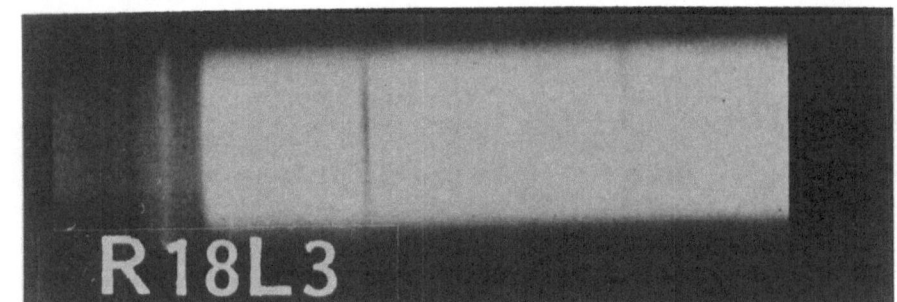

R18L3

147.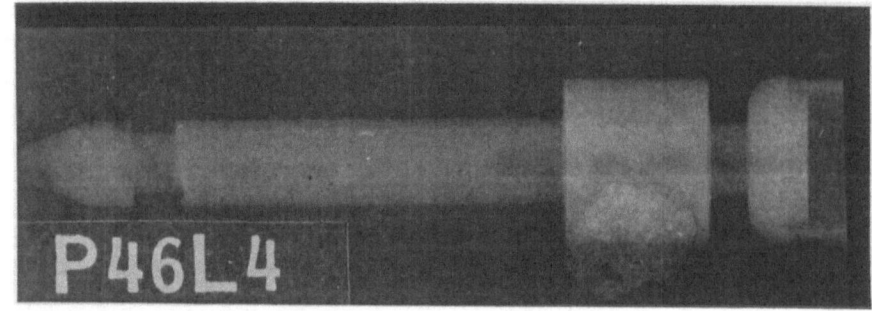

P46L4

D.PLUG a.Bottom 2.Chips

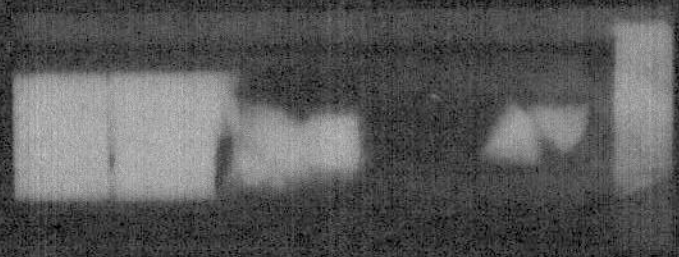

S 39L6

D.a.2.2. In plug

D.PLUG a.Bottom 6.Corrosion

R18L3

D.a.6.1. Hydrides

D.Plug b.Top 0.As fabricated

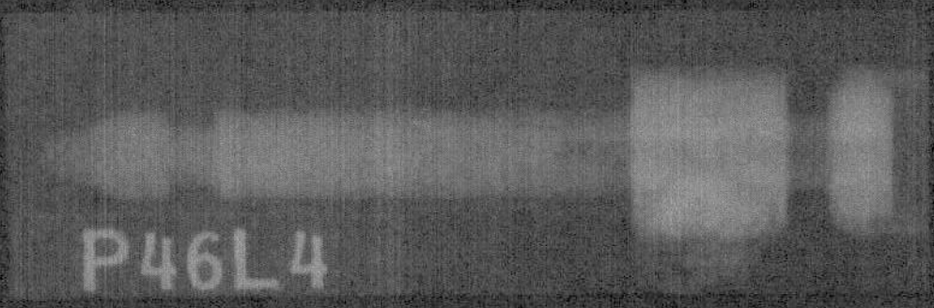

P46L4

D.b.0. (test fuel rod)

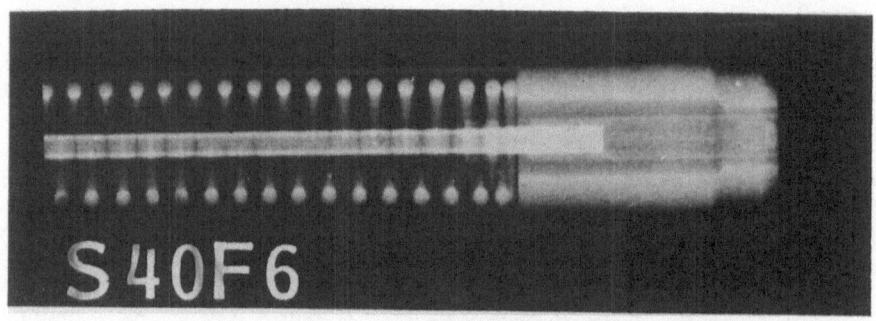

148.

S40F6

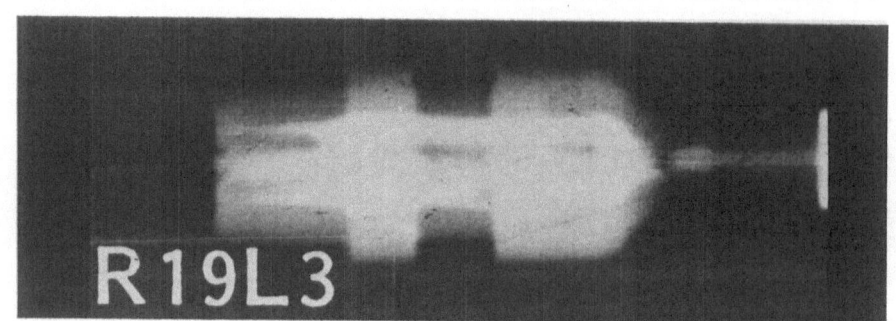

149.

R19L3

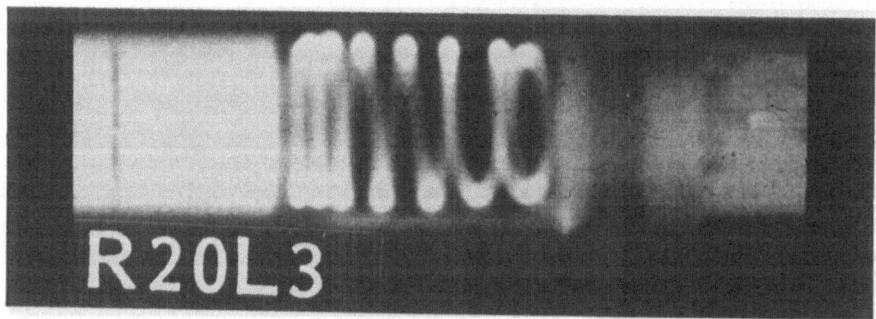

150.

R20L3

D.PLUG b.Top 0.As fabricated

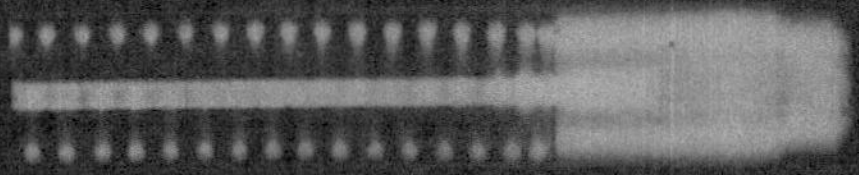

S40F6

D.b.0. (with thermocouple)

D.PLUG b.Top 3.Change of shape or location

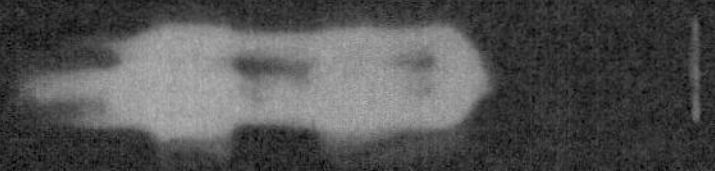

R19L3

D.b.3.5. Broken

D.PLUG b.Top 6.Corrosion

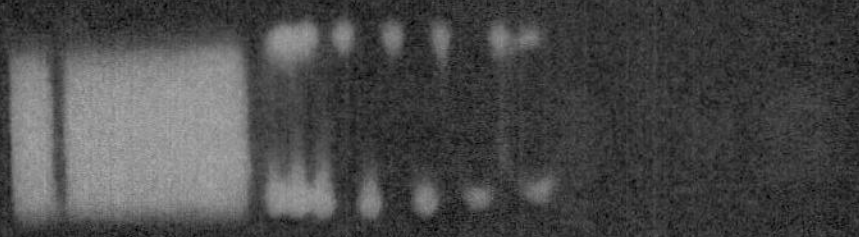

R20L3

D.b.6.1. Hydrides

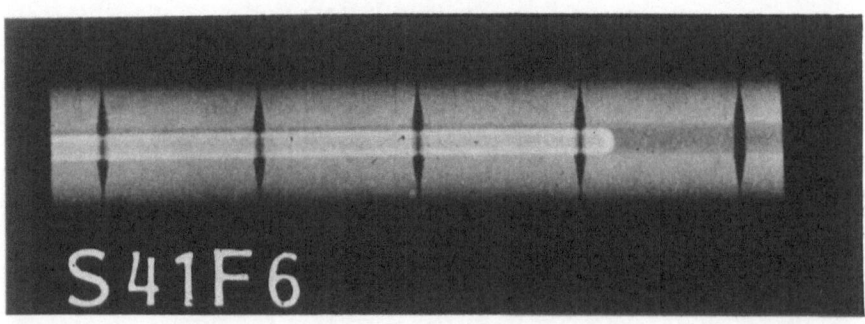

151. S41F6

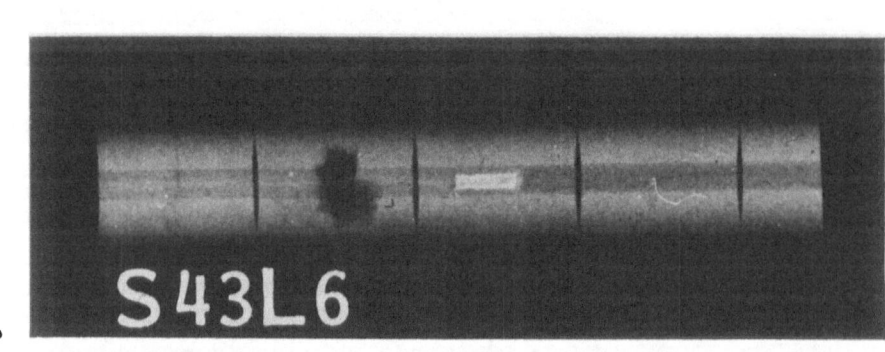

152. P47F6

153. S43L6

E.INSTRUMENTATION a.Thermocouple 0.As fabricated

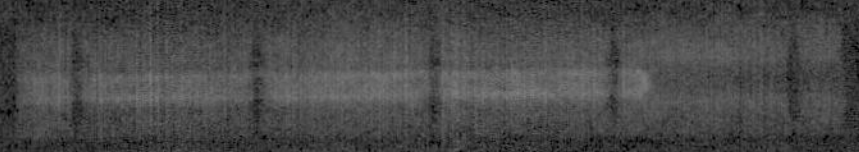

S41F6

E.a.0.

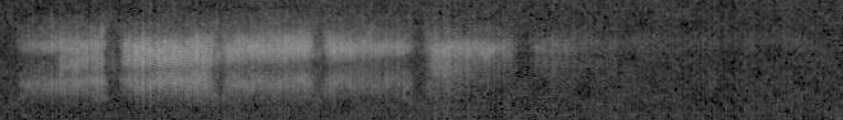

P47F6

E.a.0.

E.INSTRUMENTATION a.Thermocouple 3.Change of shape or location

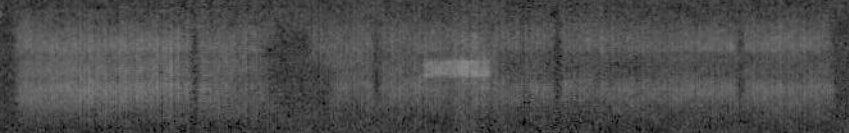

S43L6

E.a.3.10. Melted

154.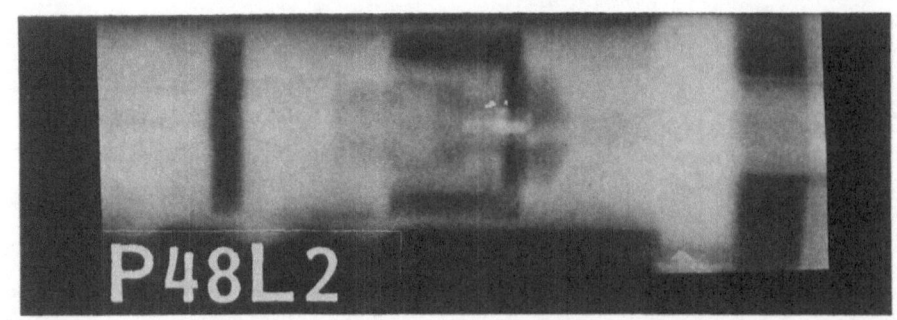

P48L2

155. G 40L2

156.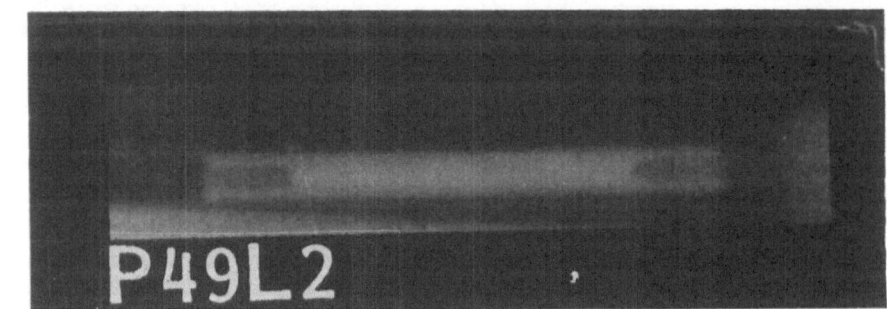

P49L2

E.INSTRUMENTATION b.Pressure transducer 0.As fabricated

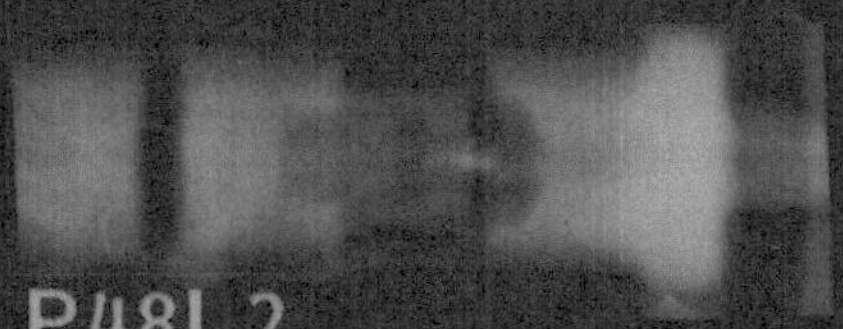

P48L2

E.b.0.   (membrane pressure transducer)

E.INSTRUMENTATION c.Diameter gauge 0.As fabricated

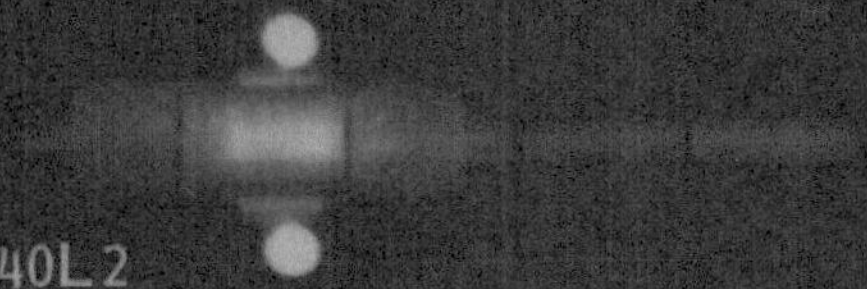

G 40L2

E.c.0.   (strain gauge sensor)

E.INSTRUMENTATION d.Length gauge 0.As fabricated

P49L2

E.d.0.   (only core)

157.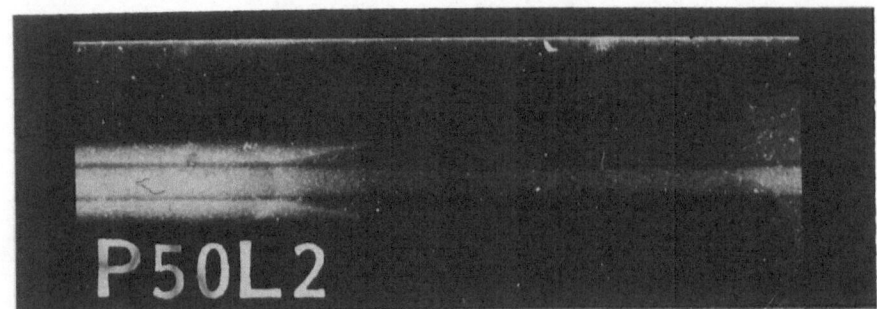

P50L2

158.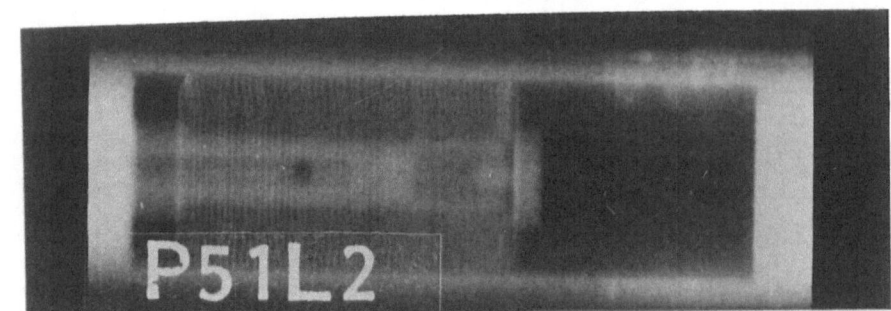

P51L2

E.INSTRUMENTATION e.Other 0.As fabricated

P50L2

E.e.0.   (Zr/stainless steel joint)

P51L2

E.e.0.   (belows system for void volume measurement)